LIGHTING

A Design Centre Publication
by Derek Phillips FRIBA FillumES MCD MArch

Note: all dimensions given in this book are in metric measurement

First published in 1966 by Macdonald & Co (Publishers) Ltd
in association with the Council of Industrial Design
2nd impression 1966, 3rd impression 1967
New and revised edition 1970

Contents

4 Introduction

6 1. An approach to home lighting

12 Visual effect of lighting

20 2. Amount of light

20 Artificial light

21 Daylight

24 3. Simple calculations

30 4. Lighting in the home

30 Fixed use spaces

34 Multiple use spaces

38 Exterior spaces

40 5. Comfort

43 Glare recommendations

43 6. Lighting Equipment

55 Lamps

59 Appendix A. Costs of lighting

60 Appendix B. Wiring and Switching

64 Appendix C. List of Manufacturers

64 Appendix D. Acknowledgements

Appendix E. List of Designers

Introduction

Life can be influenced so much by the lighting in a home. It can be made safer and easier by effective lighting, and it can be enriched by beautiful lighting.

It is hardly possible to over-emphasise the fundamental importance of lighting in the home; without light there is no 'visual' experience of our environment. We get a limited idea of the spaces we inhabit from our other senses of touch, smell or hearing, but until light is added, information of our environment is minimal. Our impression of any form is determined by the way light is reflected from it; the texture of cloth can be emphasised or subdued, a colour can be enhanced or degraded. A space can be changed dramatically by lighting. It can be 'warmed' or 'cooled'; it can be made to appear higher or lower; it can be unified or fragmented.

An Adequate Wiring System

Good home lighting is entirely dependent upon an adequate supply of lighting points, placed not necessarily at the traditional centre point of the ceiling but at strategic places chosen to facilitate the requisite 'amount' of light of the correct 'distribution', at the 'position' where light is needed. If the opportunity is there, good lighting becomes much easier so that the number and location of all lighting points should be discussed at the drawing board stage when 'extra' will be easy and inexpensive to provide.

Responsibility for Lighting

Who is responsible for lighting the home? In the first place the architect, or if there is no architect, the builder. This is a direct responsibility to provide the means whereby a house can be lit, in terms of adequate circuits and electrical points. Where the design calls for light to be co-ordinated with other structural elements, and built into recesses in ceilings, cupboards and elsewhere, the responsibility lies again with those putting up the building. The home owner starts where this leaves off (or the interior designer if one exists). He will choose the 'light fittings' which are either portable or easily fixed and changed, such as surface mounted ceiling fittings, pendants, and wall brackets.
If the owner sells the house it is the latter which he will generally

take with him and the former which, like built-in cupboards and equipment, are sold with the house. There is a definite financial advantage to be gained from ensuring that the specification for this type of 'built-in' lighting is high, and prospective house purchasers should look out for the highest standards in the lighting and wiring which 'goes with the house'.

A Functional Approach

The purely technical approach leads to a satisfaction of lighting levels to the exclusion of appearance; there is a tendency to consider lighting fittings from considerations of output only, a desire to see uniform levels of illumination to the exclusion of quality, and an insensitivity to the decorative requirements of the home.

The sentimental approach has even more obvious pitfalls; lighting fittings are chosen because they look pleasing and fit with the particular decorative approach adopted for the interior, irrespective of whether they give light of the distribution needed. The number and design of the vast majority of home lighting fittings and shades is evidence of the importance paid to this decorative aspect at the expense of efficiency. A truly functional approach is one where both the practical aspects of vision and the decorative needs of the space are taken into account.

Purpose of Book

There is ample evidence to suggest that the architect provides the 'framework for living', but that the vagaries of fashion and style dictate that much of the decision taking is done later by the owner, for those items of home lighting which do not form a part of the 'in-built fittings'. This book is designed to assist the architect to provide a satisfactory framework and the owner to achieve the environment he desires within this context.

Summary of Book

The popular approach whereby each of the rooms in the house is taken separately and analysed to offer individual solutions is tedious and unrewarding, as many remarks which apply to one space are equally appropriate to another. Here the problem is dealt with by analysing the information required from light and how this can be achieved . . . spaces themselves being divided up on the basis of whether they are such as to be solved by a single lighting solution, as in a kitchen, or whether the multiplicity of use of the space demands a much more flexible approach.

Perhaps one of the most difficult problems is to visualise the effect of a light fitting after it has been brought back from the shop and installed in a room. To help in visualising this effect, a series of photographs illustrate the light and shade which can be expected in a room from a variety of light distributions, derived from different standard lighting methods available.

Now that we are coming to expect good lighting at work, poor home lighting is very much less tolerable; and whilst one would not expect every man to be his own lighting engineer, one can very well expect some interest in lighting standards and simple calculation guides. In view of this the latest recommended levels of lighting in the home are outlined and sufficient data given to enable the wattage to provide these levels to be approximated.

One of the weaknesses in any book on home lighting is in finding suitable illustrations. There is a tendency for architects to prefer their buildings photographed by daylight, a condemnation by implication of the unsatisfactory nature of the artificial lighting; and where photographs of interiors at night are taken, the effect of the fitting is often supplemented by photo floods and other photographic techniques which render them valueless other than to judge the appearance of the lighting 'fittings' themselves placed in their setting. Specialist photography is required to obtain the real effect that the fittings produce in the room and even when this is available the absence of the subtleties of colour provides another serious weakness.

The photographs included have been carefully selected to ensure an accurate record of the type of lighting depicted.

1: An Approach to Home Lighting

There is a need for a 'functional' approach which is fundamental to all the different requirements of a home. Fundamental to:

1. Ease of performance of the various tasks that must be done, or economy of effort. Light for work, sewing, washing up, cooking, etc.
2. Safety, emphasis of danger spots in plan and the avoidance of the numerous accidents in the house, stairs, etc. (also electrical safety).
3. Economy of electrical energy. Light should not be wasted.
4. Delineation of the basic architectural framework. Sequence of spaces. Surfacing materials. Changes in ceiling height.
5. Conveying the specific decorative impression desired, e.g. applied texture/surfaces/wallpaper/pictures/furniture/colours/carpets.

An approach which is sufficiently simple to embrace all these different requirements is to consider lighting first and foremost in terms of the information it conveys to the eye. There are many sophisticated aspects of this, but these too can benefit from being considered in this simple way. For example, if too much general light is thrown on to the front of a tapestry it will appear to be very flat . . . to overcome this a higher degree of light must be added from a close angle to the surface to render the individual textures of the weave discernible to the eye. In other words too much light with the wrong geometry may serve to lessen the information received.

Any object or feature can be treated in the same way—take a step in a corridor. By daylight this may easily be negotiated, but after dark it can become a dangerous hazard, particularly to old people. Light must reveal the location and height of the step, not only when a person is adapted to a normal level of artificial light, but also when night adapted. How often we have to move about a house in the middle of the night, when lights designed for normal adaptation merely serve to blind the eye. It is simple and inexpensive to cater for this need so that parents and children can traverse any necessary part of the house with ease and safety. It is this sort of thinking which should be applied to every part of the house. Each house is different and there is, therefore, no standard solution, only standard aims and criteria. The eye adapts itself to the general level of light in a room and this may be seen easily by example.

During the day if the electric light is switched on in a well 'day lit' space, it will be barely discernible, whilst after dark the light may be more than adequate and appears to give a comfortable impression. Alternatively, if when asleep the same light is switched on, the effect on the dark-adapted eye will be such as to produce temporary blindness, until the eye has adapted itself to the higher level.

Lighting criteria should be related, therefore, to three basic adaptation levels.

1. Night adaptation level
2. Normal electric lighting levels after dark
3. Daylight.

Night Adaptation Level

The lowest adaptation level is that when the normal artificial lighting is out and the household is asleep. The level of adaptation when the eyes are opened will depend upon whether there is moonlight or whether it is very dark outside. If curtains are left open, moonlight will give sufficient light to the night-adapted eye for safe movement, for the child to go to the bathroom, for the parent to go to a child 7 who cries. On dark nights or when curtains are drawn, the natural light will be insufficient and it is desirable to place a night circuit in position at strategic points in the house to provide an equivalent of the moonlight level. This can be done easily by the use of 5 watt neon lamps set perhaps into other lighting units but separately switched, or a modern equivalent would be a small photo-luminescent panel available in America. Such lighting could be left on permanently since it would cost only a fraction of a penny to run. (5×5 watt lamps could be run continually for £1 per year).

During the day or even under normal artificial lighting levels the night circuit would give no visible light, when dark-adapted everyone in the house benefits. Suitable places would be in halls, bathroom, staircase, landing and living room.

When children are young, bedroom doors will generally be left open to ensure that a disturbed child is heard and plenty of light will be available in this way, particularly if the rooms have 'borrowed' light panels.

Where bedroom doors are shut, then for the same degree of safety a night circuit light should be capable of being switched from the bed position. This would enable a person to get up without destroying his night adaptation—a most suitable type of fitting in a bedroom would be the skirting light developed for cinemas which directs light to the floor and is shielded to prevent any view of the light source.

Simple louvred light
Rotaflex 'Framelight"

Artificial Lighting Adaptation

The illumination levels recommended for artificial lighting in the home, as listed in Chapter two, are calculated to ensure that the adaptation of the individual is well able to cope with the range of brightness when moving from one part of the house to the other.

There is a tradition that whenever a light is not needed it is automatically switched off: this tradition dies hard. It is not the case everywhere and particularly in countries where electric power is relatively cheap—in America and elsewhere, lights all over the house are switched on when required and remain on until the family go to bed. (I have personal memories of this as long as twenty years ago in Australia.) Clearly if lights are left on, the adaptation to artificial lighting levels is greatly simplified. Danger occurs when an occupant of a brightly lit room comes out on to an unlit hall or staircase. A good case can be made for leaving a sufficient number of lights on around a house for the sheer joy of experiencing a sequence of lit spaces without the irritation of switching lights on and off—but the case becomes compelling if the aspect of safety is also considered.

The architecture itself may suggest places where lighting fittings can be incorporated. Alternatively the plan should suggest points at which 'information' is going to be needed and artificial lighting points must be provided wherever the specific requirement demands.

Daylight Adaptation

This is principally the area of responsibility of the architect or builder. Windows are structural items, costly to put in or change when a building is complete. We have moved a long way since the days of the window tax and the present danger is more one of 'overlighting' our houses than underlighting by day. Windows let in and out other things than light, heat for example and privacy—and the largest windows aren't necessarily the 'optimum' size. Windows must be carefully planned to achieve 'quality' of light and shade in a building, and it will generally be found that in large rooms it is better to have more than one window and preferably in different walls, rather than the single popular picture window. Windows can be glaring, due to harsh contrast between the brightness of the window and the wall into which it is set, and this can be lessened by light from other windows at right angles or on the opposite side of the room.

The design of windows is a problem in itself. It is concerned with orientation and view, the general external appearance of the house, but principally with the quality of the light admitted—it is not something that the householder will generally determine, but it is his to control, and the use of venetian blinds and other controls such as shutters and blinds can modify the daylit appearance of spaces. An aspect which is often overlooked is the effect upon daylight of the reflectance of interior surfaces in a room. Dark wall surfaces and particularly dark floors and ceilings will reduce the level of light in the parts of a room furthest from the window and can produce glaring conditions when seen in relation to the brightness of daylight outside.

In some circumstances daylight can produce dangerous conditions. Shadows caused by sunlight can fall across a staircase in such a manner as to obscure the information of the relationship of tread to riser to such a degree that particular care must be exercised to prevent stumbling.

Daylight is a changing light source. This is both its strength and its weakness. Its strength, because change is necessary and adds a vital element to appearance; but its weakness too, because there are many days when the daylight level indoors in normally well lit rooms falls below an acceptable standard. At this point, use can be made of artificial sources; but even on dull days, the eye is adapted to the window brightness and this will be very much higher than the level of adaptation normal in artificial light, even if the recommended standards have been

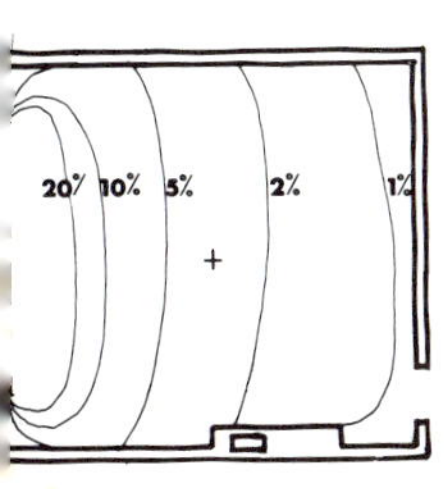

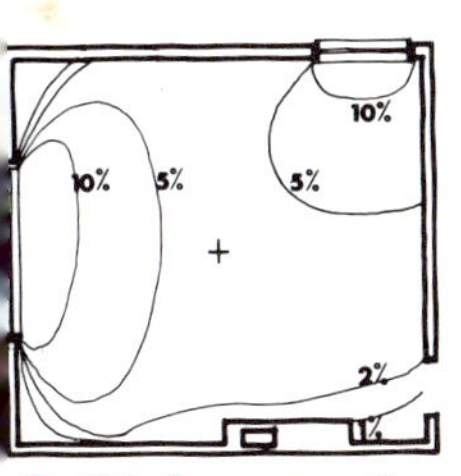

Daylight factor curves for similar rooms with different window arrangements. The centre of each room has daylight factor of 4%

9

used. For this reason it will be desirable to adopt a system of lighting which gives more light than the normal night time level.

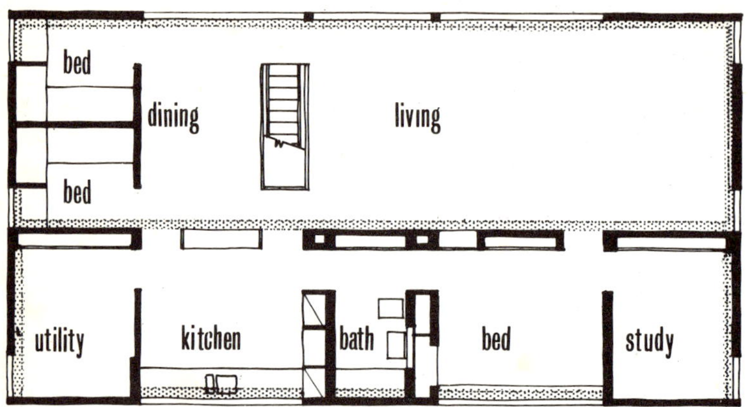

Floor plan of house at Harpenden by Jack Bonnington. The fluorescent lighting troughs are completely integrated with the organisation of the plan. The troughs run at the junction of ceiling with outside wall and are also related to the central spine wall.

Other places where a consideration of daylight adaptation is relevant exist where there is no daylight penetration. For example, a totally artificially lit bathroom. More concentrated land use in cities makes this form of planning not only possible but mandatory, to free the perimeters of building for living accommodation; in such circumstances a level of illumination more closely associated with daylight adaptation will be necessary if a dull appearance is not to be experienced during the day. Our experience of brightness will be gained from the brightness of wall surfaces so it will be helpful if any wall facing the door can be lit to a high brightness to avoid a dim impression on entering. After dark this level can well be reduced. Simple methods of achieving this are given in the latter part of the book, when methods of control and circuits are discussed.

When the planning of buildings makes it impossible to obtain well lit landings and corridors, it will help if light can be borrowed from the rooms adjacent by having panels of glass either above or at the side of doors. This can considerably increase the level of light in the circulation area. Vertical panels at the side of doors will be found generally to give the most light.

Planning Lighting

When planning the lighting of a house, the effect of these three adaptation levels must be borne in mind, for satisfactory results at all times of day.

The fundamental aspects of home lighting have already been listed and these can be related to these three levels of adaptation.

1. Ease of performance

This is principally, though not entirely, related to the 'amount' of light available.

Other factors include the direction from which light comes, source colour, and the adaptation of the eye.

2. Safety

If accidents are to be avoided information must be available at all times of particular points of danger in a home, stairs and changes of level being obvious examples, but are not alone. Others include boiler plant and cooking stoves, stores and external areas at night.

3. Economy of electrical energy

Lighting fittings should be reasonably efficient if the energy put in (watts) is not to be wasted. Clearly some types of lighting will be more efficient than others. For example, fittings giving an indirect distribution will be less efficient in terms of providing light on the page of a book than a fitting giving direct light, but the indirect fitting must be judged also on its ability to give the visual effect desired. The important thing to ensure is that the fitting is as efficient as its distribution permits.

4. Delineating the basic architectural framework

There are no easy rules here. Each house will have its own particular problems, the character of which will determine the way light sources should be related to the structure. With open plan houses areas may be defined by windows during the day; to emphasise the sequence of spaces at night, much can be done by artificial lighting to assist in defining individual areas for different uses. The placing of artificial sources should receive the same careful and initial thought as the placing of the windows.

5. Conveying the specific decorative impression desired

The need to ensure that the objects, surfaces and colours of the decorations of the house are satisfactorily conveyed to the eye. It will be useless to satisfy all the other factors and to fail in this, since it is by this, the standard of quality, that the lighting will be judged.

Here the eye is the only arbiter and success comes from practice and experience—experience of light sources and their characteristics of direction and colour, and what effect these have when placed in a particular juxtaposition with the form or surface to be lit. Individual problems present themselves which are as difficult as any that are found elsewhere, such as the lighting of a picture or a tapestry.

Lighting is not an isolated problem. It is related to the choice of colour schemes and tone values, and the final choice of colours should be made using the light source, filament or type of fluorescent lamp colour, by which it will be seen. Some very surprising things can happen to colours, two colours which may appear as a pleasant contrast by daylight can be rendered almost identical by some light sources, completely destroying a decorative idea.

The Visual Effect of Lighting

It is very difficult to judge the visual effect which a light fitting will make until it is brought home and installed. In general no information of this is available either from the manufacturer or the retail store, and yet decisions have to be made and light fittings purchased when seen in relation to many others competing for attention in a store. The majority of light fittings are, therefore, bought not on the effect they will produce by the light they give to the room, but on the 'light fitting' as a piece of decoration in its own right. Clearly the latter aspect has a validity but not at the expense of the main purpose of a light fitting.

The following series of pictures may help to solve this difficulty. It is clear that the results are strictly limited by space, but the idea is unlimited in scope and it is to be hoped that it may be developed further as a result of this initial experience.

The pictures are the result of an attempt to isolate for analysis the effect of different light distributions when derived from the sort of lighting methods generally adopted in the home. Many other combinations are, of course, possible.

An intelligent interpretation of the pictures will enable a judgement to be made of many other alternative situations e.g. the effect of a fitting giving a general diffusing distribution from a low suspended shade will be similar to the effect of a portable unit of the same distribution when placed at the same height. Alternatively, the combined effect of a number of different fittings may be assessed.

Distribution of Light

The category of 'distribution' of a fitting is based on an estimate of the relative amount of light directed up and down and is an engineering assessment. The term 'general diffusing' ·is capable of some confusion unless this is borne in mind, since the category includes both fittings which direct light in all directions such as photograph 11, and fittings with opaque sides which allow approximately equal light both up and down (12). The effect on the eye of these two fittings will, of course, be very different, although the light distribution pattern in the room will be similar. Some of the most useful fittings for general lighting in a space are of this type, the opacity of the sides being varied to give a degree of sideways brightness calculated not to cause glare.

It is possible to use some fittings to achieve a variety of distributions, a simple example of which is the opaque cylinder wall bracket which can be turned up one way to give an 'indirect' distribution to the ceiling and when reversed, a 'direct' distribution downwards.

Similarly some floor standards are designed to be particularly adaptable by providing a combination of lamps and switching to alter the balance of upward and downward light, to achieve a variety of distributions. The choice of a lighting fitting should be made first on its light distribution, and there is a clear duty on the part of the manufacturer and the retailer to ensure that this information is stated for prospective purchasers. The information is simple to provide and once a clear lead is given by the more reputable manufacturers, the custom could easily become established.

The standard classification of distribution is used throughout this book and is a widely accepted and easily understood method which should be adopted.

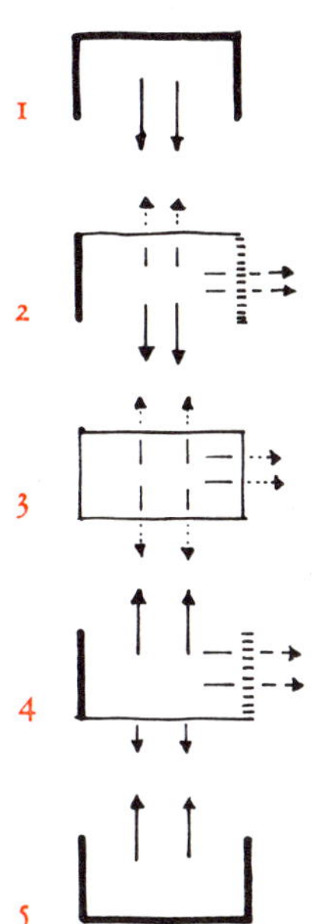

Light distribution | Definitions
1 Direct — more than 90% downward light
2 Semi-direct — 60%—90% light downwards, remainder up
3 General diffusing — approx. equal quantities of light, up and down
4 Semi-Indirect — 60%—90% light upwards, remainder down
5 Indirect — more than 90% upward light

Lighting methods

A   Ceiling mounted fittings
B   Suspended or pendant fittings
C   Wall brackets
D   Ceiling recessed units
E   Portable fittings. Floor or table lamps

The setting

The dimension and physical characterisation of the test situation are shown. All light save that from the light fitting shown was eliminated during photography.

Tabulated information

The type of fitting associated with its distribution is indicated against each picture, together with illumination levels read at points a, b, and c, shown on the plan at table level of 460 mm. The lamps and wattage to each fitting are stated. Where fluorescent lamps are used the lower efficiency 'de luxe warm' colours are adopted as being suitable for most home situations. All illumination levels given in lux (lumens per square metre).

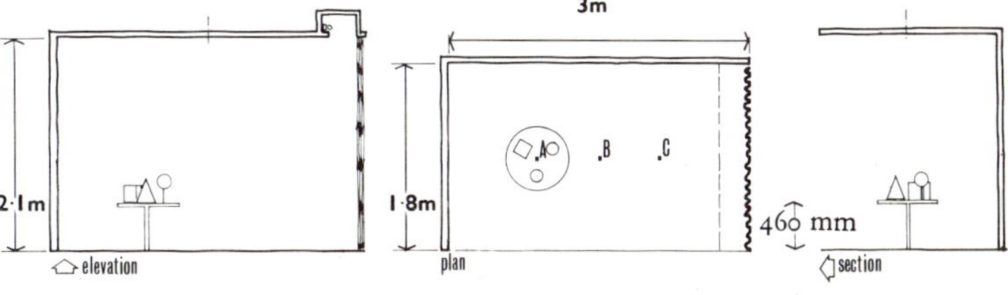

1 Direct

Method:	A. Ceiling mounted
Fitting:	Low voltage spotlight
Lamp:	1 50w, 12v internally silvered reflector lamp
Illum. levels:	a 807 b 130 c 33
Comments:	Reflected light from table almost equals direct light on top of ball–striations on wall. Use mainly for high lighting special displays.

2 Direct

Method:	C. Wall bracket 1372 mm height
Fitting:	Opaque shade-light out of bottom only
Lamp:	1 100w
Illum. levels:	a 83 b 183 c 71
Comments:	Very pronounced wall shadowing and pronounced modelling of objects.

3 Direct

Method:	D. Recessed
Fitting:	Black hole downlight with strict cut off
Lamp:	150w internally silvered reflector lamp
Illum. levels:	a 420 b 915 c 183
Comments:	Very restricted light output. Pronounced downward shadowing. Eliminates the impression of a lighting fitting. No glare. See shadow of table.

Direct **4**

Method: D. Recessed
Fitting: Fluorescent lamps
               above opal louvres
Lamps: 2 1200 mm 40w
               fluorescent
Illum. levels: a 258 b 355 c 205
Comments: Very even lighting.
               Ceiling dark by
               contrast to fitting.
               Shadows are gently
               modelled.

Direct **5**

Method: D. Recessed
Fitting: Fluorescent lamp
               recessed at junction,
               between ceiling and
               wall (in photograph a
               curtain)
Lamp: 1 1500 mm 80w
               fluorescent
Illum. levels: a 10 b 17 c 62
Comments: Lamp is placed 300 mm
               from curtain,
               enabling light to flash
               to the floor level.

Direct **6**

Method: E. Portable
Fitting: Linear table lamp
Lamp: 1 525 mm fluorescent
               13w
Illum. levels: a 161 b 24 c 3
Comments: Very localised light.
               Care needed with
               both filament or
               fluorescent table
               lamps to ensure that
               bare lamps are not
               seen.

7 Semi-direct

Method:	A. Ceiling mounted
Fitting:	Opal glass set off ceiling by dark surround
Lamps:	2 60w filament
Illum. levels:	a 101 b 118 c 74
Comments:	Reflections on ceiling inevitable, extent depending on the gloss of the paint. Good general distribution of light—tendency for fittings to appear bright.

8 Semi-direct

Method:	B. Suspended 1700 mm from floor
Fitting:	Opaque cylinder with majority of light downwards
Lamp:	1 100w
Illum. levels	a 77 b 150 c 50
Comments:	Good general lighting. Local ceiling brightness above fitting. Distribution depends upon suspension height.

9 Semi-direct

Method:	C. Wall bracket 1372 mm height
Fitting:	Linear wall bracket with opaque front
Lamp:	1 80w 1500 mm fluorescent
Illum. levels:	a 108 b 118 c 86
Comments:	Limited wall pattern due to lamp's close proximity to surface—tends to emphasise imperfections in wall surface.

Semi-direct 10

Method: E. Portable

Fitting: Floor standard with light reflected downward but with some light upward.

Lamps: 2 60w filament

Illum. levels: a 237 b 323 c 205

Comments: Use of shade tends to convert into general diffusing fitting, but reflector incorporated throws more than 50% of light downwards.

General diffusing 11

Method: B. Suspended 1143 mm from floor

Fitting: Spherical luminous shade

Lamp: 100w filament

Illum. levels: a 54 b 280 c 52

Comments: Effect is of generally even light, soft modelling. The larger the globe for the size of lamps, the less the possibility of glare.

General diffusing 12

Method: B. Suspended 1788 mm from floor

Fitting: Opaque sided fluorescent fitting with opal louvres

Lamps: 2 1500 mm 80w fluorescent

Illum. levels: a 409 b 570 c 302

Comments: Good general light. Less pronounced modelling than more direct suspended fittings.

13 General diffusing
Method: C. Wall bracket
              1372 mm height
Fitting: Opal glass giving
              light all round
Lamp: 1 100w filament
Illum. levels: a 108   b 118   c 86
Comments: The light immediately
              below the fitting is
              reduced by the
              support.
              Approximately
              equal light is dis-
              tributed all round the
              fitting.

14 General diffusing
Method: C. Wall bracket
              1372 mm height
Fitting: Opaque cylinder with
              light up and down
Lamp: 1 100w filament
Illum. levels: a 100   b 205   c 85
Comments: Approximate 50% up
              and 50% down
              categorises this as
              generally diffusing.
              But the opaque sides
              reduce the brightness
              towards the eye.

15 General diffusing
Method: E. Portable
Fitting: Linear fluorescent
              directed to wall
Lamp: 1 525 mm 13w
              fluorescent
Illum. levels: a 14   b 12   c 9·7
Comments: Fitting screens bare
              lamp from the eye.
              Light limited to one
              side. Similar to (14)
              but with light source
              away from wall
              gives good general
              distribution to wall.

Method: B. Suspended
Fitting: Pendant translucent
            glass bowl
Lamps: 2 100w filament
Illum. levels: a 302   b 344   c 23
Comments: Very traditional
            method, tends to
            produce flat
            modelling.

Method: C. Wall bracket
            1372 mm height
Fitting: Opaque cylinder
            with light up only
Lamp: 1 100w filament
Illum. levels: a 50   b 50   c 43
Comments: Pleasant subdued
            effect. Reduced
            modelling. Pro-
            nounced wall pattern.

Method: E. Portable
Fitting: Reflector upwards
            only—surrounded by
            translucent shade
Lamp: 100w filament
Illum. levels: a 86   b 89   c 69
Comments: Similar effect to 17
            without pronounced
            wall pattern.
            Wall patterning
            reduced by edge
            lighting of shade.

2: Amount of Light

Guidance on Home Lighting is contained in the following publications:

1. *British Standards Code of Practice CP 68/133454 General Series 'The Provision of Electric Lighting in Dwellings'* available from the British Standards Institute (under revision).
2. *The I.E.S. Code, 'Recommendations for Lighting Building Interiors 1968'* available from the Illuminating Engineering Society.
3. *Interior Lighting Design Handbook* (April 1969. Electricity Council, £1).
4. *Home Lighting Initial Planning.* Lighting Information Sheet No. 25, published by the Electricity Council. Some information is also available from the Ministry of Housing and Local Government in the Housing Manual, 1949, and its Technical Appendices, and their later publication 'Homes for Today and Tomorrow', available from H.M. Stationery Office.

The Code of Practice recommends the wattage for different light sources required to light rooms in the home, but as these figures are based on illumination levels considered satisfactory in 1948, they are an inadequate basis for the higher illumination levels recommended today. It is understood that this Code of Practice is now under revisision and a new issue is likely to become available (C.P. 68/33454).

The latest recommendations on the amount of light required throughout the home are tabled in the Code of the Illuminating Engineering Society, published in 1968. These levels have been adopted as a standard throughout this book, although it should be borne in mind that these are recommended minimum levels for people with average eyesight, and that higher levels are required for the aged.

General advice on lighting, and accurate lighting calculation is contained in the Electricity Council book 'Interior Lighting Design Handbook'.

Artificial Lighting

A simple method of calculation is outlined in the next chapter but as units of light are not so widely known and understood as units of heat or electricity it may be helpful to give some general information on illumination levels.

The unit of light flux is the lumen, and illumination is measured in terms of lumens per sq. metre (lux). To put this in actual terms, daylight outside in average conditions will provide 5,400-10,800 lux. In areas of bright sunlight levels of 32,000 will be found. On a moonlit night the illumination level will be in the order of ·18 lux and in starlight as little as 1/10 of this amount. The adaptation of the eyes permits us to see a lesser or greater degree in these extreme conditions.

Recommended Illumination Levels in the Home

	lux (lumens per square metre)
Living rooms: General	100
Reading (casual)	200
Sewing and darning	600
Studies (desk and prolonged reading)	400
Bedrooms: General (supplementary local lighting should be provided at mirrors)	50
Bed-head	200
Kitchens (working areas)	200
Bathrooms (supplementary local lighting should be provided at mirrors)	100
Halls and landings	100
Stairs (at tread)	50–100
(fittings should screen the lamps from view both when going up and down the stairs)	
Workshop (benches)	400
Garages	50

An important distinction is made in these recommendations between rooms of fixed use, kitchens and bathroom, for which a specific amount of light is required and other rooms where the amount of light is related to the job to be done. This emphasises the flexibility required in the latter areas of the home, a flexibility to which the artificial lighting must contribute a positive role.

The increase in the levels of light required in the home is in line with the overall increase in artificial light levels in other fields. This is right. Many jobs performed in the home are no less exacting than similar jobs in the office or in industry and people who have adequate lighting at work are unlikely to be satisfied for long with less in their home.

In America even higher levels are encouraged and this is probably related to the fact that the more you have the more you want, which seems to apply also to other aspects of environment, such as heating.

Daylighting

Only a proportion of the daylight available outside enters a room, and this is reduced the further one gets from the window. As daylight varies, the amount of daylight in a room can only be expressed in terms of the ratio between the inside and the outside.

This ratio is known as the daylight factor and is expressed as a percentage of the total light available from a complete hemisphere of sky, excluding sunlight. In a typical situation in a house, the variation in daylight between the front and the back of a room may be from 9% down to 1·3% which means that with average conditions outside (10,800 lux) the fall off in illumination will be from 9% of 10,800 to 1·3% of 10,800 or from 970 lux to 140 lux.

This means that in rooms with windows which meet Ministry requirements the amount of light close to the window during the day will be sufficient for all likely needs, but that towards the rear of spaces some tasks such as sewing and darning will be insufficiently lit, and this is common experience. We tend to move closer to the window when faced with difficult seeing tasks.

At night we cannot hope at present to provide levels of light which are equivalent to those from daylight close to the window, but we can equal and even improve on levels of daylight in areas of the house far from windows. An indication of the sort of illumination levels derived from different light sources is given in the sketches below.

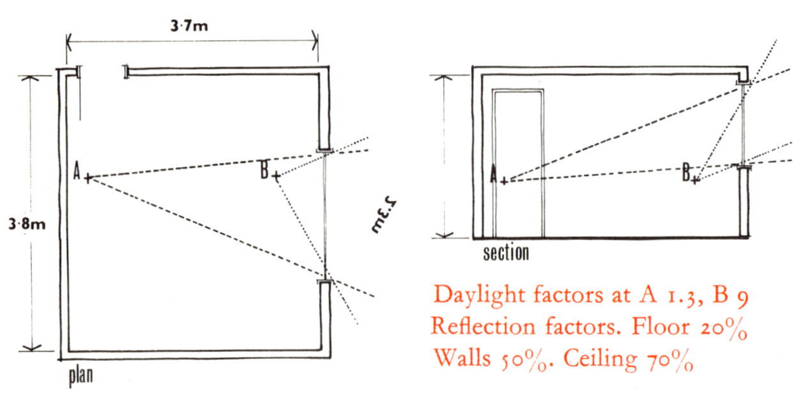

Daylight factors at A 1.3, B 9
Reflection factors. Floor 20%
Walls 50%. Ceiling 70%

60 watt adjustable fitting
A 485 lux under lamp
B 258 lux (300 mm) from A

100 watt reflector lamp
Person A has 860 lux on reading surface
Person B has 645 lux on flat table surface

Uniformity of Light

It is unnecessary, and in most circumstances undesirable, to have completely uniform lighting. This tends to give a dull overall appearance and reduces shadows and modelling, the principal characteristics of 'information' of three dimensional objects.

Uniform lighting is only required when the task is 'uniform' and for example may be necessary in some offices or factories. In the home this is unlikely to be the case and when 750 lux is recommended for sewing or darning this does not mean that a level of 750 lux must be provided uniformly throughout the room . . . it does, however, require that the general level of illumination is not too low, so that harsh contrasts of brightness are avoided in the field of view, a subject dealt with later in the chapter on 'Comfort'.

Control of Amount of Light

The important thing about 'amount' of light is that, whilst sufficient must be available when it is required, for much of the time the family may be sitting, relaxing or watching television when the illumination level can be related to the minimum 'information' for safe movement in the room. The accent should therefore be on flexibility of lighting both in the amount of light made available and in its location in the room.

The lighting solution must be capable of providing the required amount of light at the required place in the room and the lighting fittings should be planned with this in mind. In rooms of a single use a single solution, but with rooms of multiple use a considerable flexibility of control. Methods of control by switching and by means of dimmer circuits are dealt with in Appendix b, but as a general guide the solution will be made easier the more lighting positions and points there are available: it is better to have a number of sources of small wattage than to rely on a single point (e.g. the central ceiling point) of high wattage.

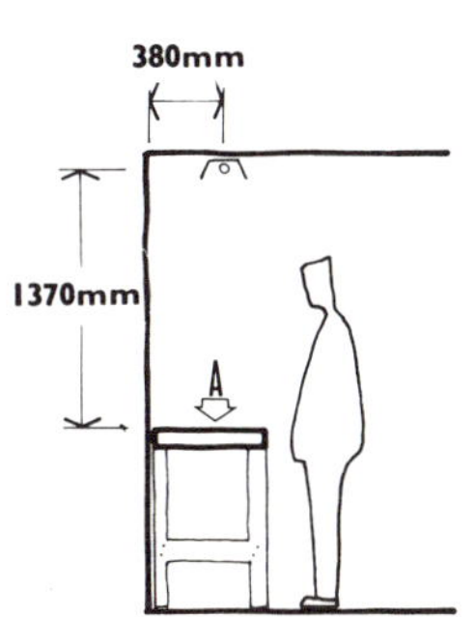

◀ A 1500 mm 80 watt fluorescent fitting with opal plastic reflector mounted over a workbench.
A. Under lamp 430 lux

3 × 60 watt lamps in fitting 1200 mm ▶ high 610 mm from fitting person A has 222 lux on reading surface. 1200 mm from fitting person B has 64 lux on reading surface.

3: Simple Calculation of Wattage Required in Rooms

It must be emphasised at the outset that the method of calculation described is only a broad approximation. It does not attempt to provide really accurate solutions*. It is rare in home lighting that really accurate solutions will be required, and the method suggested may be used to check whether the lighting planned is adequate, either for a new house, or when changes are contemplated to bring the standard of lighting in an old house up to date.

Method of Calculation:

Before starting, a list of relevant facts should be compiled.
   The purpose of the room with a list of the activities likely to be carried out.
2. The dimensions of the room. A rough plan should be drawn to enable the room area to be worked out. Having the plan drawn out is useful for placing the light fittings later.
3. Decorations—whether the wall surfaces or curtains will be light or dark.

Procedure

1. Look up the table of recommended illumination levels in Chapter 2 and select for the activities already listed.
2. Select the light source to be used, e.g. fluorescent or filament, or a combination of both.
3. Determine the distribution of light from the light fittings to be used. (Clearly less wattage will be needed if highly efficient fittings are employed. The five basic types of distribution already outlined (Chapter 1) are used in the following table and it will be apparent that least wattage is required using direct fittings with the fluorescent source, and most wattage when filament lamps are used in indirect pelmets or cornices.)

*Accurate methods of calculation are available and are adequately described in such publications as *Interior Lighting Design Handbook* published by the Electricity Council.

Use the following table as a guide to determine the lamp wattage required for typical domestic interiors. with room heights of 2.4 m.

Light distribution of Fitting	Approx. lamp wattage (w) required per sq metre of room to give one lux in service		
Direct	A	B	C
Filament	·35	·29	·25
Fluorescent	·12	·10	·08
Semi-Direct			
Filament	·37	·33	·27
Fluorescent	·12	·10	·08
General Diffusing			
Filament	·39	·34	·28
Fluorescent	·13	·11	·09
Semi-Indirect			
Filament	·42	·36	·29
Fluorescent	·14	·12	·10
Indirect			
Filament	·69	·61	·50
Fluorescent	·23	·20	·17

Use Column A for rooms below 11 m² in area
        B for rooms between 11 m²–23 m² in area
        C for rooms above 23 m² in area.

The value of W given for fluorescent lamps are for the lower efficiency types with good colour properties recommended for domestic use. For high efficiency commercial lamps, the values of W can be multiplied by ·6 giving a lower wattage requirement.

The figures are worked out for average reflections from decorative surfaces—if decorative surfaces are very light or very dark adjustment in required wattage can be made. The reflectances of decorative surfaces used in the calculation are as follows:

Ceilings   70%
Walls   30%
Floors   15%

Notes for using the table

1. From the information on room size already noted, select the appropriate column of figures, A B or C and choose the appropriate figure for W allowing for the light source and light distribution of fitting to be used.

2. To use the table adopt the formula:
Wattage required $= w \times$ illumination level required $\times$ floor area of room

Examples of simple calculation

Example 1: A kitchen 3 m $\times$ 3·5 m

    Recommended Illumination level   —   200 lux
    Light distribution of fitting   —   direct recessed fitting
    Light source   —   fluorescent

Using the formula, selecting from Column A
    wattage required $= ·12 \times 200 \times (3 \times 3·5) = 252$ watts

If, on the other hand, a filament lamp was used the wattage required $=$ $·35 \times 200 \times (3 \times 3·5) = 735$ watts, an indication of the fact that even medium efficiency fluorescent lamps are nearly three times as efficient as filament lamps of equivalent wattage.

Example 2: A small bathroom 2 m $\times$ 3 m

    Recommended Illumination level   —   100 lux
    Light distribution of fitting   —   general diffusing
    Light source   —   filament

Using the formula selecting from Column A—wattage required
        $·39 \times 100 \times (3 \times 2) = 234$ watts.

The above illustrations have been purposely drawn from situations where a single lighting solution can be used, and the placing and spacing of the fittings presents little problem.

Distribution of Light

The spacing of fittings required to give even illumination over an area is related to their distribution, and 'direct' fittings, particularly those which are deeply recessed, have a more limited distribution than general diffusing and indirect fittings. With the comparatively low ceiling heights in modern homes, it is unwise to space recessed fittings much wider than 1200–1500 mm apart if reasonably even illumination is required. In a kitchen, however, provided that the light gives the required information at work points in the space—over cookers, working surfaces and sinks—there can be a reasonable fall off in light elsewhere.

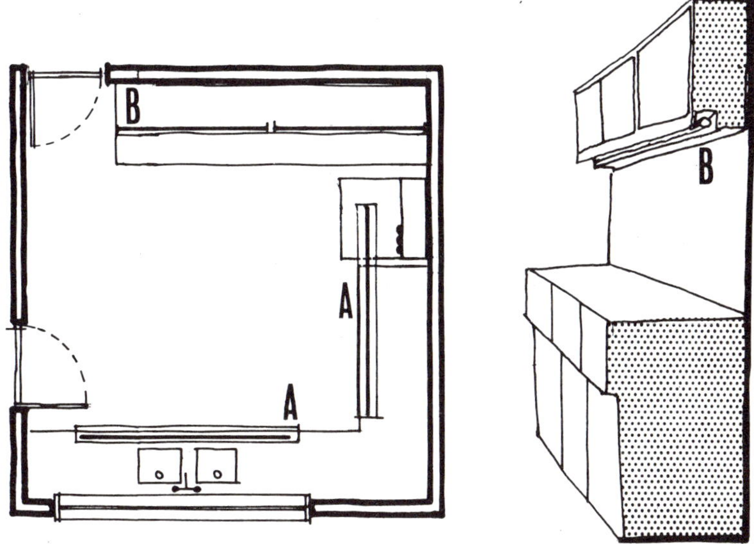

Kitchen

A 1500 mm 80 watt recessed fluorescent fittings

B 1200 mm 40 watt fluorescent fittings under high level cupboards

The above drawing indicates a suitable layout for lighting the kitchen shown in the example 1. It should be noted that purely on a wattage basis two 2400 mm fluorescent lamps (125 watts each) would be considered satisfactory, but this is ruled out when spacing and placing fittings are investigated. It would be difficult to place the two 2400 mm lamps in a location where they would provide light to all the important areas of the room, and at some points it would entail working in one's own shadow.

Calculations for rooms of multiple use are complicated by the fact that a variety of illumination levels will be needed, in addition to which it is probable that light fittings of different distributions will be used so that the calculation should be made once for each type of distribution, having made a preliminary estimate of the proportion of light required from the different distributions.

Example 3: A living room, size 7 m × 5 m

Recommended Illumination levels:
Sewing and darning   —   600 lux
Homework   —   400 lux
Reading (casual)   —   200 lux

Light Distribution of Fitting:
30–40 per cent of light from indirect sources
60–70 per cent of light from semi-direct sources

Light source:
fluorescent for indirect light
filament for semi-direct sources.

It is clear that an illumination level of 600 lux is not required over the whole room and that it will be possible to locate a light fitting to provide this as required—but when 600 lux is used, the average level of light should not fall below 1/10th of this figure for comfortable seeing. In this case a figure of 60 lux seems reasonable for the initial calculation.

If 30–40 per cent of 60 lux is to be provided by fluorescent then this represents 20 lux. This leaves 40 lux from filament sources.

Calculation (a) selecting from column C
    wattage from indirect/fluorescent light =

$$\cdot17 \times 20 \times (7 \times 5) = 119 \text{ watts.}$$

Calculation (b) selecting from column C
    wattage for semi-direct sources/filament =

$$\cdot27 \times 40 \times (7 \times 5) = 378 \text{ watts.}$$

To simplify, say 120 watts indirect/fluorescent 400 watts semi-direct/filament. A lighting solution following these lines might be planned as follows:

Solution 120 *watts of fluorescent*
    3 × 1200 mm 40 watt lamps   (e)
    400 *watts of filament*
    3 × 100 watts/ceiling fittings   (a)
    1 × 60 watts/wall bracket at desk   (b)
    1 × 40 watts/table lamp.   (d)

If the room is planned in this manner the resulting illumination level will be sufficient background for additional sources providing the higher level required for difficult seeing tasks; these might be handled by floor standards of a decorative or functional type, in which the lamps are of sufficient intensity to obtain the highest illumination levels required (see Chapter 2) at a reasonable distance from the lamp (c). In this case a direct type of fitting using a 150 watt lamp would be ample. Consideration should be given to the use of dimmers to enable a greater variety of lighting effects and intensities to be achieved.

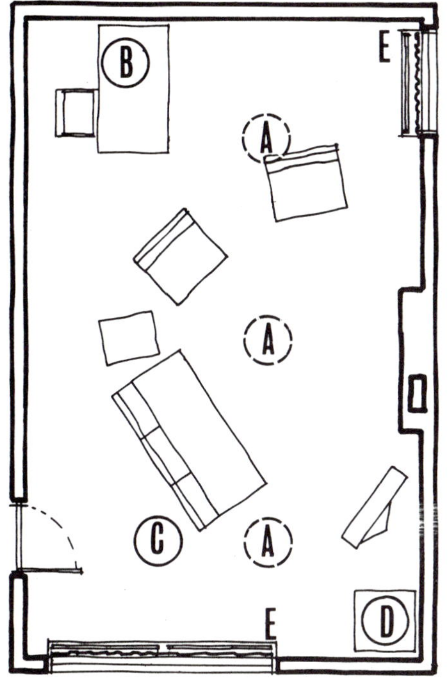

Living Room
A   3 × 100 watts/ceiling fittings
B   1 × 60 watts/wall bracket at desk
C   1 × 150 watts/floor standard for occasional use
D   1 × 40 watts/table lamp
E   3 × 1200 mm 40 watts fluorescent lamps

Summary

Calculation of this sort can be a useful check on the amount of light which should be made available; it has the advantage of simplicity, enabling individuals, without a great deal of trouble, to obtain a measure of guidance. This guidance is limited to 'wattage' and it must be associated with the consideration of 'spacing' to avoid unbalanced lighting, inadequate in parts and overlit elsewhere.

Other criteria such as electrical distribution and switching, interior decoration and flexibility must be considered and the final lighting solution will evolve from a commonsense approach to all these considerations. In the past an overemphasis has been placed on the decorative aspect of the light fitting itself; fittings have tended to be chosen for their appearance rather than their performance. Whilst 'appearance' both lit and unlit is a valid consideration, the 'performance' of a light fitting contributes much more to 'living in the home'.

4: Lighting in the Home— A Discussion of Problems

Discussion of Actual Examples

The lighting in a home resolves itself into three types of problem, the solution to all of which may be treated with a good deal of variety according to the decorative style of the home, its architectural characteristics and the individual needs of its occupants.

Three types of home lighting problem:

(a) Fixed use spaces, or areas in which a static solution is adequate, since the uses of the space do not change, generally associated with fixed items of furniture.

These will include laundry, kitchen, utility rooms, workshop, garage, bathroom, w.c., circulation areas, corridors, halls and staircases, landings, storage, cupboards, larders.

(b) Multiple use areas, or areas in which a variety of different activities take place, either at the same or at different times, or where family living calls for a variety of approaches to the same activity. These areas are generally associated with a flexibility of furniture layout.

These will include living room, dining room, combined kitchen-dining room, combined living-dining rooms, bedrooms and nurseries, family playrooms.

(c) Exterior Areas or areas outside for which consideration should be given after dark. These will include, the garden, patio or barbecue, paths and driveways, porches, service yards and fuel stores, swimming pool.

a. Fixed Use Spaces

These present the simplest problem and a review of homes throughout the country indicates that more adequate solutions are to be found to it than b or c. A simple method of calculation has already been described and this is sufficient to enable an adequate electrical loading (wattage) for the space to be assessed. After this the actual layout and choice of individual light fittings can be made, bearing in mind that it is the 'information' we need which must result . . . whether this is information of the crockery in the sink to minimise breakages and ensure hygiene, information of the form of a staircase to ensure safety, or information leading to a positive appreciation of a picture in the hall, or the sequence of spaces resulting from the architect's organisation of the house plan.

Fixed use spaces are associated generally with fixed elements, the sink, cooker and work surfaces in the kitchen, laundry equipment, the workbench in the workshop (whether part of the garage or separately located), the bath and lavatory fixtures in the bathroom and the staircase and doorways in circulation areas.

It will be clear where most light is wanted and if the overall wattage for the space is adequate the location of light fittings in relationship to areas where the most information is required will generally ensure adequate levels over the rest of the space.

Staircases are a particular instance of a lighting problem where the distribution of light must be such as to ensure accurate and unequivocal information to the brain. Here the 'amount' of light should be adequate but it is its 'geometry' or the manner of modelling which it provides which is of the first importance.

To walk safely up a staircase, there should be a clear distinction between stair tread and stair riser and this can most easily be got by arranging that there is a light source located at the top of the staircase which will brighten each of the nosings to the treads, but will throw a shadow on the risers. If, at the same time, a small quantity of general illumination is provided from the front of the stairs a reasonable balance of brightness will be struck.

Where a landing is provided in a staircase which changes direction, the fitting which provides the general illumination to the top flight can be chosen to have a semi-direct distribution so that its main light will provide the necessary directional light to the lower half of the staircase.

It is attention of this sort to staircases or at changes of level in plan, where one or two steps may be necessary, that will increase the safety of the home.

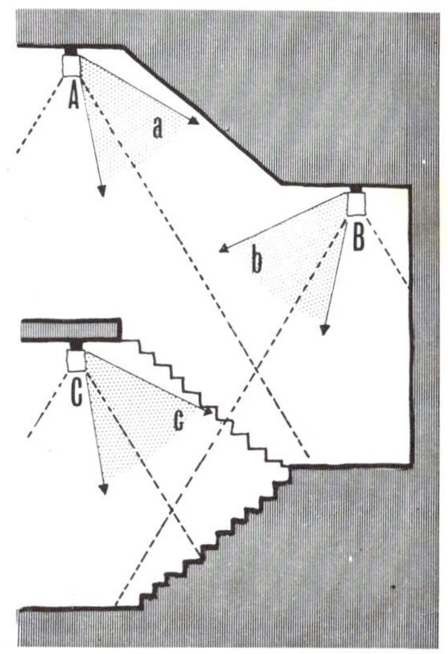

Staircase. Example of light for 'information'

A Direct light, illuminates upper stair nosings, shadows upper stair risers

a Diffused light, illuminates wall surfaces

B Direct light, illuminates lower stair nosings, shadows lower stair risers

b Diffused light, illuminates upper stair risers, reducing harsh shadowing

C Direct light, illuminates lower stair access

c Diffused light, illuminates lower stair risers, reducing harsh shadowing

1 Storage unit. House at Harpenden.
Fluorescent lamps wall mounted behind cupboard units to give indirect distribution to shelving and back lighting to television unit. In rooms without fireplace such a storage unit can act as a focal point when lit. Provided that the wiring is provided initially, this type of solution is inexpensive and would be considered as an asset when selling the house.
Architect: Jack Bonnington
Photographer: Dennis Hooker

2 Kitchen. House at Harpenden.
The general lighting is provided by recessed fluorescent lamps placed along the window wall, with additional fluorescent light above the cooker and at the built-in breakfast bar in the foreground. The system relies for its efficiency on the reflective capacity of the white Venetian blind slats. High illumination levels are achieved at all the important work areas. All lighting designed and built-in as an integral part of the house design.
Architect: Jack Bonnington
Photographer: Dennis Hooker

3 Kitchen. House at Bovingdon.
Emphasis on central working surface and sink unit using fluorescent lamps above small cell louvres. General illumination from 6 100 watt recessed downlights gives light to cupboards and window unit. All light sources integral with design.
Architect: Derek Phillips Associates
Manufacturers: Concord Lighting International
          Downlights, Ecloplas louvres
Photographer: Jewell-Harrison

4 Bathroom: House at Edgware.
A 1500 mm 80 watt fluorescent lamp behind a wall mounted pelmet with a general diffusing distribution gives very adequate lighting. (Using the method of calculation outlined in Chapter 3, example 2; if fluorescent is substituted for the filament lamp, it will be found that 1 1500 mm 80 watt lamp gives the required 100 lux in a 6 m² bathroom.) This form of lighting is integral with the building.
Architects: Challen and Floyd
Photographer: Dennis Hooker

5 Internal bathroom. House at Harpenden.
This small bathroom, which is internally ventilated, has a roof light to assist the general illumination during the day. The artificial lighting comes from a recessed 1500 mm 80 watt fluorescent lamp at the junction of ceiling and wall. At night the laylight can incorporate artificial sources to raise the illumination level at the mirror. All lighting integral with the house design.
Architect: Jack Bonnington
Manufacturer: British Lighting Industries Ltd
Photographer: Dennis Hooker

6 Bathroom at a Design Centre, London, exhibition.
Fluorescent lamps used above a luminous ceiling formed of opal polystyrene louvres giving a direct downward distribution. With this method a good light can be obtained without glare, but there is a possibility of looking up through the louvres to the lamps when lying back in the bath. By having a double circuit of lamps a high and low illumination level can be obtained to suit daylight and night-time adaptation for internal bathrooms. A 5 watt neon lamp placed above could act as a night circuit. This method is relatively expensive. It conceals the true ceiling and can be applied to existing bathrooms of sufficient height. Steam may be extracted through the louvres. This form of lighting would normally be the architect's responsibility and be sold as a part of the house structure.
Manufacturer: Elco Plastics—Elcoplas ceiling

3

4

5

6

b. Multiple Use Spaces

These present the greatest difficulty since no one lighting solution will be adequate for the changes of use, changes of furniture layout and changes of atmosphere which must be catered for.

The problem has been introduced in example 3 in Chapter three, where in the calculations for a living room the need for variety of solution has been emphasised. It seems unnecessary to stress the need for flexibility of lighting in a living room where the wide changes of use and occupant makes 'variety of effect' of paramount importance. It is technically possible to place a luminous ceiling over the living room designed to provide the highest amount of light for the most difficult seeing task, but such a static solution to such a complex problem would be unlikely to win approval for longer than it takes for the novelty value to wear off—it would become a bore, and it would almost certainly be substituted by more sympathetic methods in which shadows and pools of light play a part.

Variety of solution is clearly a question of economics. It would be ideal if it were possible to provide, at the touch of a switch or other control, a variety of illumination level, colour and distribution of light, at any point desired in a room. This is not an impossible technical problem, but it is sufficiently impossible economically as not to be worth discussing at any length in this book. The question is 'how much can we afford' and whilst this will vary with financial ability, a discussion of feasible solutions may be helpful.

Distribution of light

The effect of 'light distribution' is illustrated in Chapter one and clearly fittings which permit a number of alternative distributions offer an increased range of possible room appearance. In some cases this may be a matter of altering the position of the fitting, e.g. portables, extended arm fittings; in others by changing the way the fitting is pointing, e.g. adjustable spots, reversible wall brackets, and in others it may be a combination of lamps which varied by the switching, provide a variety of distribution.

The light distribution can be varied by the number and variety of fittings which are turned on and the emphasis must be on increasing the possible flexibility of effect by ensuring adequate lighting points, for the different lighting methods available, e.g. surface or pendant recessed and concealed wall bracket or portables.

Amount of light

The number of fittings turned on gives control of 'amount', but there are other possibilities which should not be ignored.

Some fittings provide alternative switching so that the wattage varies to give different illumination levels; similarly alterations in distribution of individual fittings will contribute to changes of general illumination level.

A more adequate method is by dimmer control. Dimmers are now available for dimming either one or more light fittings from the switch position, so that a scale of intensity from maximum down to zero can be obtained.

Dimmers are of use in altering intensities in 'fixed use' spaces used at different times of day. For example the lighting of an internal bathroom could be changed to cater for three levels in association with daylight, normal artificial light or alternatively night time adaptation.

Colour of light

There are not at present any cheap and simple ways to vary colour. There are, of course, coloured filament lamps and a variety of different colours of fluorescent lamp. Methods of colour changing using fluorescent lamps are available but expensive.

A solution to the problem of colour variation is by incorporating different coloured lamps within individual lighting methods, easy to do in areas of luminous ceiling or luminous wall, but more difficult in the individual suspended or portable fitting. This enables different colours to be obtained by switching a different 'sequence' of lamps.

1 Living room/dining room associated with kitchen.
   House in Hampstead.
   An approach almost directly opposite to the house in
   Harpenden, where virtually no lighting is built-in. The
   methods include pendant Noguchi paper shades over the
   dining table, an Anglepoise lamp at the desk mounted on
   the bookshelf system and two photographic reflectors
   with 100 watt lamps which can be directed up or down.
   Considerable variety of appearance can be achieved with
   this simple means. Apart from the necessary overhead
   wiring for the pendants, the remaining wiring is provided
   from socket outlets at skirting level, the photographic
   reflectors being mounted on a timber batten fed from
   below. All these lighting fittings are easily removed.
   Architect: Michael Brawne
   Manufacturer: Herbert Terry & Sons Ltd, Anglepoise
             lamp. Photographic reflectors.
   Photographer: Arnold Behr

2 Living room. House in Scotland.
   A room in which there is virtually no 'built-in' lighting.
   The lighting methods used are wall mounted adjustable
   spots or portables using filament lamps. Distributions
   vary from direct to general diffusing. A variety of
   appearance and amount of light can be achieved by
   combinations of portables and by adjusting the spots to
   light the walls, curtains or ceiling. The room has a quality
   of small pools of light and shadow at low level which is
   associated with a comfortable domestic atmosphere. The
   general illumination level is low. Portable table lamps in
   the centre of rooms create difficulties of safety of wiring.
   It is probable that all light fittings in this installation
   would be removed on change of occupant.
   Architect: Morris and Steedman
   Manufacturer: Rotaflex Lighting Ltd, Spotlights
             Portables, Continental
   Photographer: John Dewar Studios

3 Dining room. House at Eastlake, Ohio, USA.
   A pendant formed of 3 basketwork shades hung over
   the dining table, in addition a wall pelmet lighting
   painting and brickwork pattern. Both methods providing
   semi-direct distribution. In a dining room it is important
   to have 'cross lighting' to people's faces to avoid harsh
   shadows and this type of shade assists. The pendant fitting
   would go with the owner; the pelmet remain with the
   house.
   Courtesy, General Electric, USA

4 Dining room. Conversion at Bovingdon.
   The principal lighting method is by wall washer fittings
   recessed into the suspended ceiling. These light the walls
   and pictures and provide cross lighting. All wall washer
   units can be dimmed to reduce the general level for dinner

parties to enable candles to be used on the table as in this
picture. The timber servery screen has built-in tubular
filament lighting up and down from a shelf at table level
with sand blasted glass. The orange ceiling is lit from two
indirect wall brackets and appears too dark when left
unlit due to the nature of the wall lighting. It is important
to note that downlighting over the table kills the quality
of candle light and has therefore been omitted. All lighting
being built into ceiling or furniture would be regarded
as a part of the house.
Architect: Derek Phillips
Manufacturer: Rotaflex Lighting Ltd.
          Troughton & Young (Lighting) Ltd
Photographer: Gordon Giles

5 Living room. House at Harpenden.
   A very closely integrated system of linear recesses at the
   junction between wall and ceiling, in which simple
   fluorescent batten fittings are concealed. (See page 10 for
   plan of building.) Additional light to end wall by
   I. S. Reflector flood in floor standard—this can be used for
   providing a high level of illumination on work when
   required. A degree of flexibility of appearance and
   illumination level provided by switching individual lines
   of recessed lighting in association with a lit storage unit
   (see page 32) and floor standard. A solution of this sort
   must be planned with the initial structure of the building
   and becomes an integral part of it.
   Architect: Jack Bonnington
   Manufacturer: Atlas Lighting Ltd, fluorescent
   Photographer: Dennis Hooker

6 Extension to living room. House at Edgware.
   During the day the lighting of this built on living room is
   assisted by the daylight opening which gives light also to
   the original living room behind. At night by means of
   fluorescent lamps placed at the sides shielded by the
   venetian blinds, the same opening is used as a luminous
   ceiling. The remaining lighting is by table lamps and
   floor standards of a general diffusing nature. The combined
   systems give a glare-free environment with a high
   average level of lighting above 30 lm/ft². The loading is
   3.5 watts/ft². When the filament lighting is used alone, a
   level of 6 lms/ft² gives a more gentle effect with pools of
   light. On dull days the fluorescent lamps can be used to assist
   the daylighting of the internal room. With the exception of
   the daylight all the fittings are easily removable by the
   owner.
   Architect: Challen and Floyd
   Manufacturer: Rotaflex Ltd,
             Hille & Co Ltd, filament,
             Atlas Lighting Ltd, fluorescent
   Photographer: Dennis Hooker

2

3

4

5

6

c. Exterior Spaces

The prime fact to bear in mind in the lighting of exterior spaces is 'that a little light goes a long way in the dark' and that a 40 watt lamp used outside will appear, by contrast, brighter than 100 watt indoors. Exterior spaces include problems of both fixed use, and multiple use. For example, the lighting of external service areas, garden paths and driveways can be solved by a single installation, whilst some flexibility of approach and effect must be applied to the garden, the barbecue or the swimming pool. Fixed lighting points should be provided for the former, but an installation with the characteristic of addition and change is more logical for the latter.

For lighting outside at night a view of the lamp itself will almost inevitably produce a heightened adaptation and a consequent reduction in visibility, so lamps themselves should be screened or so diffused by external shades or diffusing glasses that the brightness of the filament is broken up. Where possible only the light reflected from the grass or shrubs should be visible, and a greater sense of depth will be apparent if light sources are placed at varying distances from a view point, adding perspective.

1

2

4

1 Special shield fitting used for garden path lighting. The fitting uses a 40 to 60 watt lamp and provides sideways light through louvres.
Designer: Derek Phillips and Associates
Manufacturer: Rotaflex Lighting Ltd
Photographer: Sydney Pyzan

2 Landscaping. House at Woldingham, Surrey.
External floodlights sunken below the garden level flood brick wall seen beyond the living room of the house, giving an extension of view at night. In this view the sliding windows have been removed so that there is no problem of interior brightness reflected from the glass. With the glazing in position the brightness of interior light fittings and surfaces must be controlled so that the view towards the outside wall is not marred by reflections. The wiring for exterior lighting should be planned with the building, when it will be inexpensive to install. The facilities given by this method will be a built-in asset in the sale of the **property**.
Architect: Derek Lovejoy
Manufacturer: Philips Electrical Ltd
Photographer: The Ideal Home magazine

3 Barbecue. House at Bovingdon.
Filament fittings used exclusively together with candles and the glow from the barbecue. Garden paths lighting by 40 watt lamp in a portable stalk fitting giving louvred light downwards. Shrubbery lighting by 100 Watt PAR external flood directed upwards to leaves. Traditional Spanish hanging lantern (25 W) over table with added candle brackets. This example indicates the effect of very low levels of lighting outside at night. Provided that exterior wiring points are built-in this form of lighting can be carried out easily by clients' individual fittings. These would not normally be fixed and could be brought indoors during the winter.
Architect: Derek Phillips and Associates
Manufacturer: Rotaflex Lighting Ltd, garden path
Photographer: Dennis Hooker

4 External lighting. House at Harpenden.
A combination of lines of fluorescent lamps recessed into the underside of the first floor slab screened with louvres and semi-recessed filament fittings with circular glass diffusers designed to give a semi-direct distribution to lighten the slatted wood ceiling. The system gives plenty of light for general lighting of approaches whilst the lines of fluorescent articulate the structure at junctions of glazing. The illumination builds up towards the house to assist in adaptation. This form of lighting is so essentially a part of the structure that it would certainly be sold with the house.
Architect: Jack Bonnington
Manufacturer: Atlas Lighting Ltd, fluorescent
Allom Heffer & Co Ltd, filament
Photographer: Dennis Hooker

Summary

As a general rule a light fitting should be thought of in functional terms, to provide 'information', and this can most easily be done by concealing the light source itself so that it is the 'song' not the 'singer' which is experienced. This does not rule out the decorative value of light sources for their own sake as objects in a room but where use is made of this sort of 'furniture' it is generally unwise to rely on it in any functional sense for the provision of light as 'information', except where very low levels are acceptable.

5: Comfort

Contrast

Glare is a question of contrast, and light sources which appear bright and glaring in one situation will appear almost invisible or even dark when their surroundings are changed. The simplest example of this is the headlamp of an oncoming car, which against a dark night will appear glaring, whilst if switched on during the day will be scarcely noticeable. Taking this a step further, the headlamps would even appear dark if they were seen against the higher brightness of a setting sun.

The contrast to a job of work in the home is also important, for if the brightness of the background is higher than the task it will cause discomfort—this is most difficult to overcome when the task itself is a dark colour which reflects little light.

Ideally, the background to the task should be one-third as bright as the task itself, and the general surroundings should not fall much below 1/10th of that of the task. The higher the level of illumination required, the more care should be taken to avoid glare.

Comfort

Comfort is a sensation not necessarily enjoyed when discomfort is absent, but the first task in designing for comfort is to eliminate discomfort. Conditions of comfort or discomfort are subjective and personal phenomena which cannot easily be measured; one individual's reactions will vary from those of another. Some people are unusually sensitive to glare conditions, so that conditions that will satisfy most people will prove glaring to a few, for whom special precautions for the avoidance of glare must be made. When levels of illumination are recommended for particular jobs, there is an assumption that the correct amount of light is provided 'in a glare free manner' and it is necessary to explain what causes glare, where glare is often found, and what should be done to lessen it.

Glare

There are basically two types of glare; discomfort glare and disability glare.
Discomfort glare is glare from a light which is too bright in relationship to its background, its position relative to normal directions of view and its apparent size. This type of glare does not stop a task being performed but performance is maintained at a physiological cost in terms of eye strain which can lead to headaches. Glare discomfort comes both from the size and the brightness of the source, seen in contrast to its background or surroundings and light fittings which will create discomfort against a background of dark walls or ceiling may be acceptable in lighter surroundings.
The classic case of glare discomfort is experienced when trying to read in a room in which a single light from a reading lamp or floor standard is directed on to a white book. In those circumstances the brightness of the white page of the book will be extremely high, creating an unacceptable contrast with both its immediate surround and general background. In such circumstances other general lighting is most necessary; this is easily achieved, first by allowing some light to reach the ceiling to reduce contrasts and by creating a reasonably well lit general environment; second by paying attention to advice already given to avoid any bare light sources. The most critical angle is that from the eye to 45 degrees from the downward vertical and since the home problem is one in which any part of the room may be used, the light source must be shielded to give a cut off angle of 45 degrees.

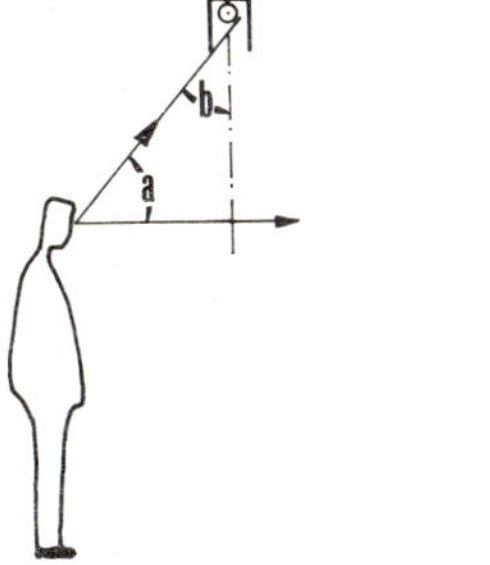

a   shielding angle
b   cut off angle—ideally
    not more than 45°

Disability glare is glare from a light source which is placed in close visual proximity to something which has to be seen, of such a brightness that it is difficult, or impossible, for a task to be performed. A typical instance of this would be trying to see any detail below a window cill— the high brightness of the window raises the adaptation level of the eye

to a point where it is impossible to see details at a much lower brightness below the cill. This effect ceases when the brightness of the wall
is raised or that of the window reduced. This is common enough and
the same effect is gained from artificial light sources placed in too
close a 'visual proximity' to the eye when trying to read.

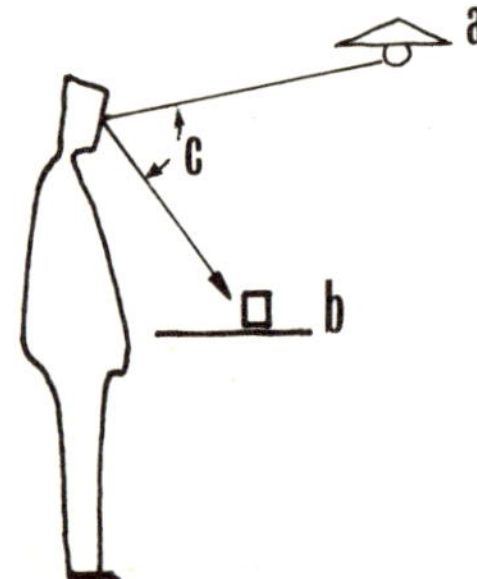

a   light source
b   task
c   angle of separation. This should
     be maximum

The two important aspects are the angle of separation and the brightness of the source. The effect is most easily eliminated by increase of
angle of separation to at least 40 degrees by raising the light fitting
out of the way of the eye (bearing in mind that discomfort may still
continue to a higher level). Alternatively, some reduction of glare will
result from a reduction in the brightness of the source either by screen-
ing it or lowering the wattage of the lamp.
Disability glare can also occur from the reflection of light sources
which are not in themselves visible. This is often found when reading
glossy magazines, the reflection of the light source tending to veil
the printed page, or reduce the contrast between print and paper to a
point when reading becomes difficult if not impossible. In these
circumstances, it is usual to turn the page away to avoid this reflection
and it is not generally too serious in a home situation, except at a desk
where the light source is fixed in such a way that reflected glare is
inevitable when books are placed flat.

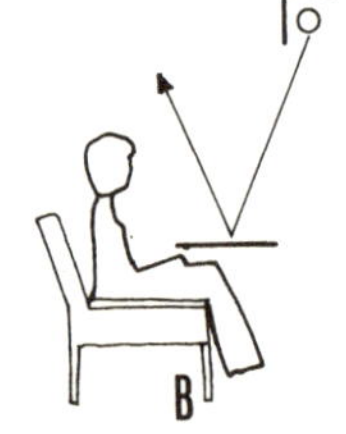

A   Person receives reflected
     glare from concealed lamp
B   Lamp has been raised.
     Glare is no longer
     reflected at person

Check on Glare in your Home

A good way to see if the lighting in your room satisfies conditions
of comfort is to see if you can read a newspaper easily . . . this is one
of the most difficult conditions to satisfy, since it requires looking

upwards, and the paper itself cuts out light from in front.

Recommendations

The Code of Practice recommends that artificial light sources should not ordinarily exceed a brightness of two candelas per square inch and on no account more than ten candelas per square inch at angles from the horizontal to 30 degrees below the horizontal when the fitting is mounted on the ceiling.

Minimum brightness recommendation: Code of Practice.
A. Controlled zones brightness should not exceed 2 candelas/inch²

This provision in the Code precludes the use of bare filament lamps and all but the lowest loaded and lowest efficiency fluorescent lamps (e.g. the 4 ft 0 in. deluxe warm white lamps 1·75 c/square inch). It is a good principle to avoid the use of bare filament lamps completely and to accept bare fluorescent lamps only of the type mentioned and only in situations where the criteria for comfort is not of prime importance, as for example in working kitchens and workshops, it should never be permitted in areas used for sitting or relaxation. Recommendations alone will not, however, produce conditions of comfort. It is an aspect of 'regulations' that they are generally of a 'negative' character and this is certainly true in the case of glare; recommendations for the elimination of glare conditions are a discipline rather than a guarantee that comfortable conditions will necessarily follow. Comfort is a combination of decorations and colour, furniture and the effect that the interplay of light makes on the whole quality of the room . . . but it is true to say that in planning for comfort, first eliminate glare.

6: Lighting Equipment

General Method of Description

Never at any time before has there been such a number of fittings offered to the public for lighting their homes, and if to this is added equipment not 'offered' in the home lighting market, but which can well be used for it, the variety is considerable.

It is clearly impossible to describe anything like the complete range of fittings, particularly if the large numbers of continental goods are included; the following is, therefore, an attempt to summarise some of the types of fittings available limiting the choice to goods produced and manufactured in Great Britain. The fittings are listed using the same method of analysis as elsewhere in this book, by lighting method and distribution.

Purchase Tax

The present method of assessing whether lighting fittings are liable to purchase tax restricts the items which manufacturers offer for home lighting. The reason for this is that all fittings offered for home lighting are subject to purchase tax. If a fitting at present used in commercial buildings or schools for which purchase tax is not payable, is offered for use in the home then not only those fittings sold by retailers for the home attract purchase tax, but the same fitting used in other buildings would automatically become liable also. This means that the fitting of one manufacturer offered in the home market would not be able to compete in other markets against an identical type of fitting by another manufacturer who does not offer it for the home, and on which purchase tax would not, therefore, be charged.
There are many fittings used in shops and stores or schools which are most suitable for the home, but which cannot be advertised for domestic use for this reason. Architects and designers who hold the catalogues of manufacturers' complete ranges of fittings, have a distinct advantage over the general public in this respect, since they can choose any suitable fitting. It is to be hoped that this anomaly which restricts the use of fittings in the home may ultimately be removed.

Advance in Techniques in Other Fields

Techniques used generally in home lighting have scarcely changed since the days of the gas lamp—but techniques in other fields have made great advances, and there are many techniques, such as the low voltage spot developed initially for shop lighting, which can be applied to the home. Another field which has a particularly interesting future in new homes is the use of simple commercial fluorescent batten fittings, incorporated in recesses in ceilings, in pelmets and in other architectural details.

Development in luminous ceilings is yet another field with application in the home. Originally designed for classroom lighting in the United States, the luminous ceiling method has been applied to nearly every type of programme and has relevance in the fixed use areas of the home —the kitchen and the bathroom being specific rooms where it has been used successfully. Considerable variety of luminous panel is available made possible by developments in the field of plastics.

Architectural Style

This book is limited in its illustrations to lighting equipment for modern homes, and does not deal, except in terms of principle, with the many, and often delightful, period interiors. Many solutions appropriate to today's architecture would be wrong if applied to an architectural style designed originally for the candle, and there are many rooms in which it would be unthinkable to use anything other than crystal chandeliers. The fact that this type of fitting is omitted from this book is not intended to suggest an exclusion of their use in the right place.

Notes on Equipment Section

1. Whilst some of the equipment illustrated is unique to the manufacturer listed, in certain cases, as with the standard fluorescent batten, similar equipment is obtainable from most equipment manufacturers. These batten fittings will rarely be visible and their choice can be confidently made on price from a reputable manufacturer.

2. The control gear associated with fluorescent fittings is normally integral with the lamp in a fitting, but where the space provided is minimal the gear can be placed remotely.

3. Equipment has been illustrated for the most common distribution but in some cases by adjustment or different fixing, the distribution may be varied, e.g. some wall brackets placed facing downwards provide a direct distribution but when inverted change to indirect.

4. The section on exterior fittings is limited, reflecting the poor selection of suitable equipment available from British manufacturers.

Direct - filament

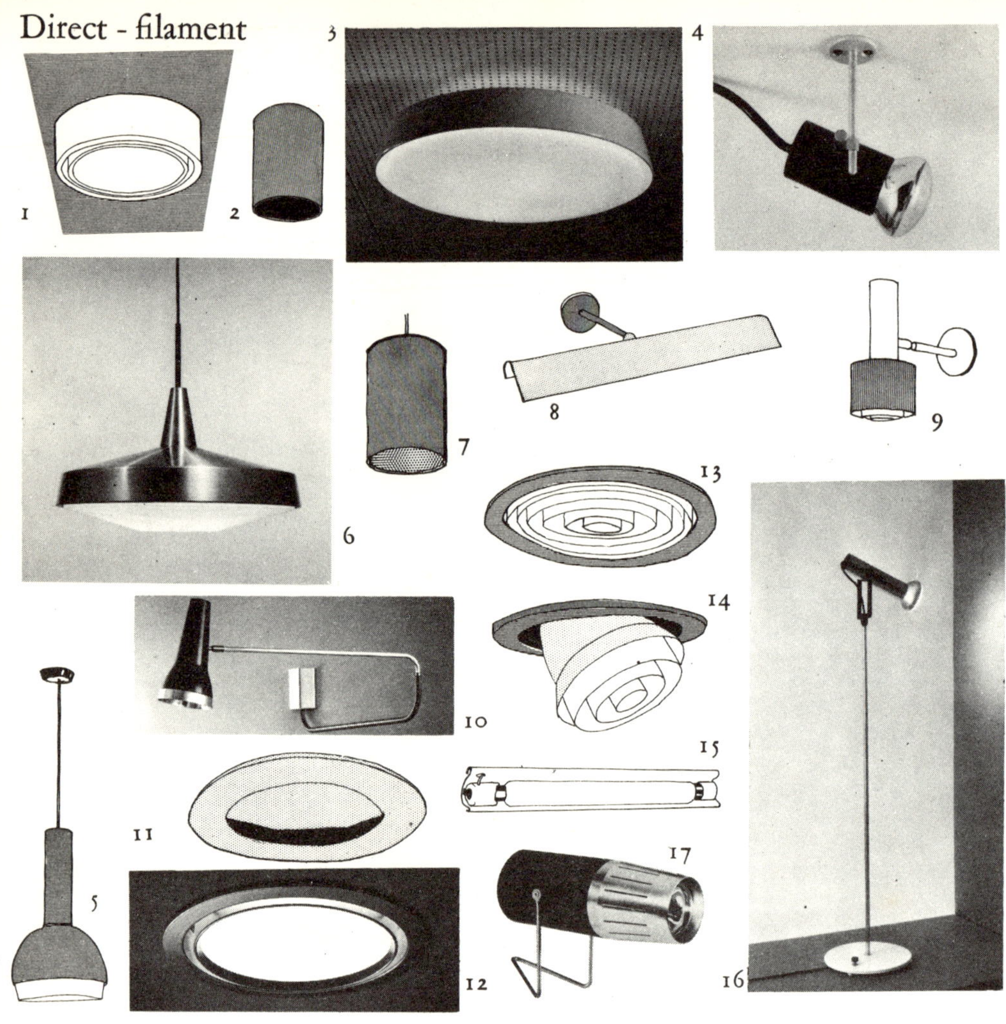

Ceiling mounted			
1 Atlas Lighting	DKF 2060	Spun metal with diffusing glass	
2 Rotaflex	8301	Spun aluminium with black baffles	
3 Merchant Adventurers	MA 1556 MH	Opal glass, metal reflector	
4 Atlas Lighting	DAS 1050	Adjustable metal holder for reflector lamp	

Suspended		
5 Troughton & Young	F695	Aluminium anodised in various colours. Rim opal glass
6 Chrysaline	MR 1	Spun aluminium. Finish copper or other colours. Moulded plastic diffuser
7 Lumitron	6211	Metal cylinder with low brightness interior

Wall mounted		
8 Merchant Adventurers	800 Series	Adjustable aluminium reflector
9 Courtney Pope	TF 80	Adjustable spotlight, anodised aluminium
10 Cone Fittings	GT/W	Anodised aluminium with coloured reflector

Recessed		
11 Rotaflex	8674	'Wall washer' fitting
12 Allom Heffer	DRG Series	Opal glass. Satin aluminium trim
13 Merchant Adventurers	1930 Series	Plastic louvre. Satin aluminium trim
14 Atlas Lighting	DS 1150	Adjustable in two directions. Metal. Painted white
15 Linolite	Reflector	Small section reflectors for tubular lamps. Easy concealment

Semi direct - filament

Portable

16	Rotaflex	4201	Satin aluminium cylinder
17	Marler Haley	Compact Spot	Low voltage spotlight with built-in transformer

Ceiling mounted

18	Allom Heffer	SG Series	Metal drum with opal glass diffuser
19	Osram (GEC)	F 36155	Opal glass with teak veneered metal cylinder

Suspended

20	Churchouse	LG 251	Grey inner cylinder with copper outer
21	Benjamin	Coolicon	Inexpensive plastic shade. White.

Wall mounted

22	Cone Fittings	AP	Washable acetate shade
23	Atlas Lighting	JYC/S1	Handwrought glass various shapes and colours

Recessed

24	Allom Heffer	CSRM	Semi-recessed opal glass cylinder. Mirror above

Portable

25	Rotaflex	3205	Standard. Perforated metal drum
26	Rotaflex	3401	Perforated metal cylinder on adjustable mounting
27	Hiscock Appleby	TL 2010	Pleated plastic shade, satin chrome metalwork

General diffusing-filament

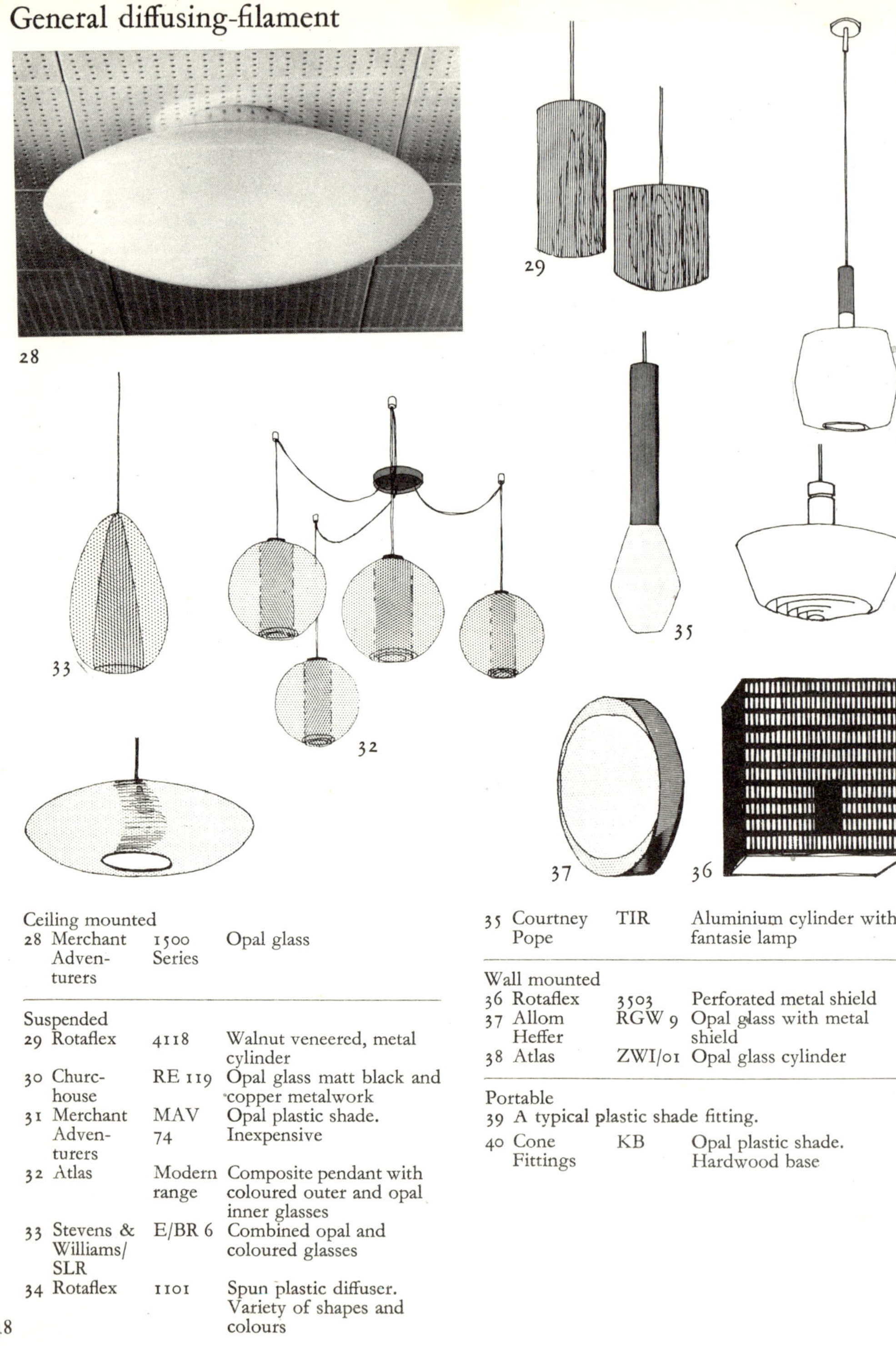

Ceiling mounted

28	Merchant Adventurers	1500 Series	Opal glass

Suspended

29	Rotaflex	4118	Walnut veneered, metal cylinder
30	Churchhouse	RE 119	Opal glass matt black and copper metalwork
31	Merchant Adventurers	MAV 74	Opal plastic shade. Inexpensive
32	Atlas	Modern range	Composite pendant with coloured outer and opal inner glasses
33	Stevens & Williams/ SLR	E/BR 6	Combined opal and coloured glasses
34	Rotaflex	1101	Spun plastic diffuser. Variety of shapes and colours

35	Courtney Pope	TIR	Aluminium cylinder with fantasie lamp

Wall mounted

36	Rotaflex	3503	Perforated metal shield
37	Allom Heffer	RGW 9	Opal glass with metal shield
38	Atlas	ZWI/01	Opal glass cylinder

Portable

39	A typical plastic shade fitting.		
40	Cone Fittings	KB	Opal plastic shade. Hardwood base

Wall mounted filament

1 Concord Lighting
International
Directional Spot
4551 Parabolic spot
Concentrated glare-
free spot/adjustable

2 Conelight
Bed-Head/General
Light
EIP cyl.
Pin-up wall light for
easy mounting related
to s/o at skirting level

3 Atlas
Decorative Glass
XW1/XFL glass

4 Merchant
Adventurers
Opaque sided up or
down
2134B/D/L/or BG
Opaque sided fitting
giving up and
downward light with
louvre, decorative
glass, or opal glass
infill

5 Concord Lighting
International
Bed-Head/Low
Voltage
Superjet 4448
Low voltage
directional spot for
reading in bed

6 Merchant
Adventurers
Indirect Light Up
2208
Scoop fitting for
upward or downward
light. Can be
provided on arm

7 Conelight
Light on an Arm
GT/W
Adjustable reflector
on swinging arm

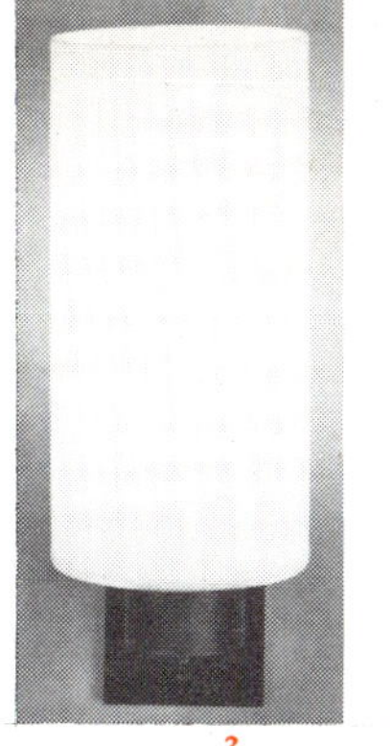

1   2   3

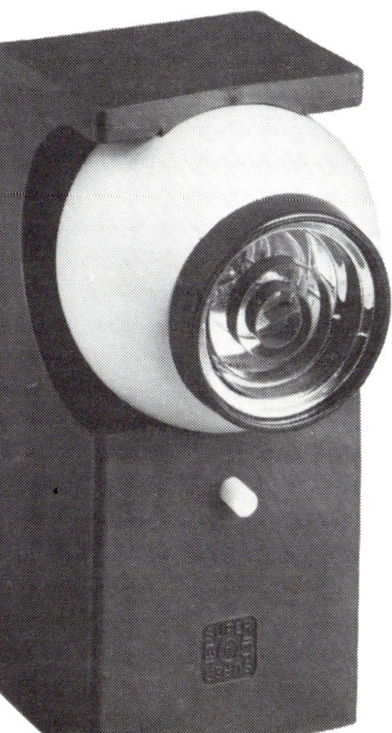

4

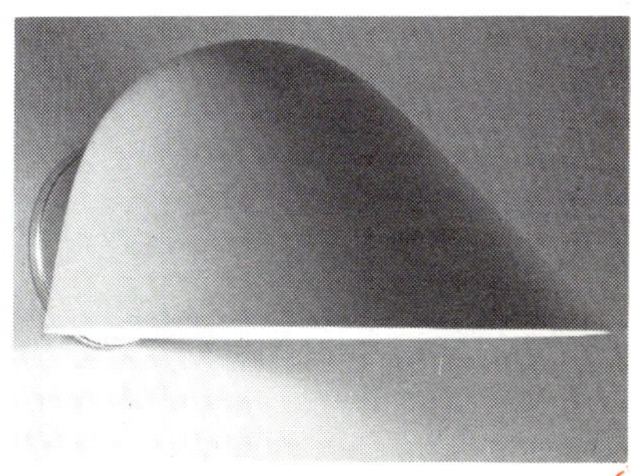

5   6

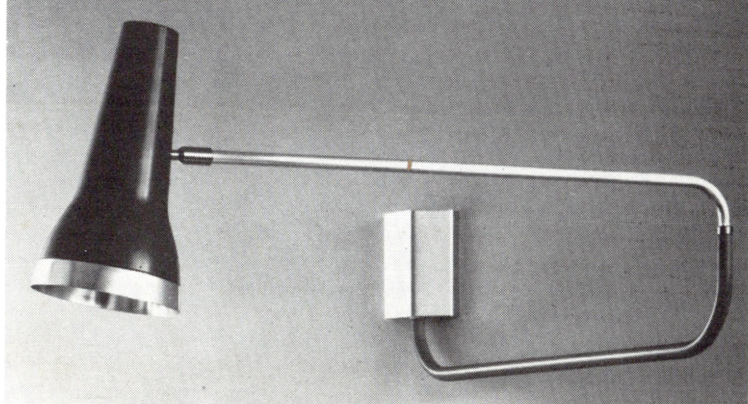

7

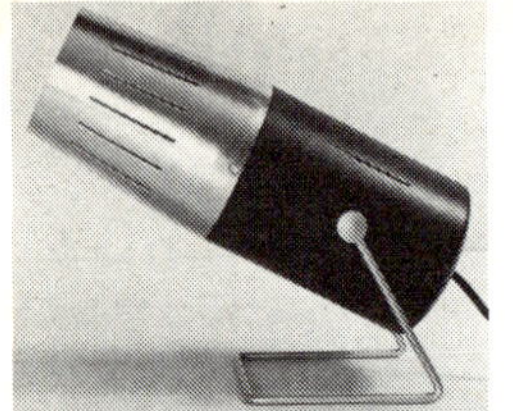

1

2

3

4

5

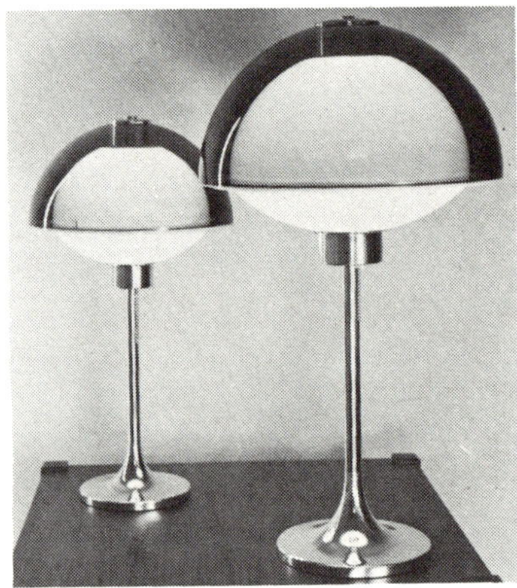

6

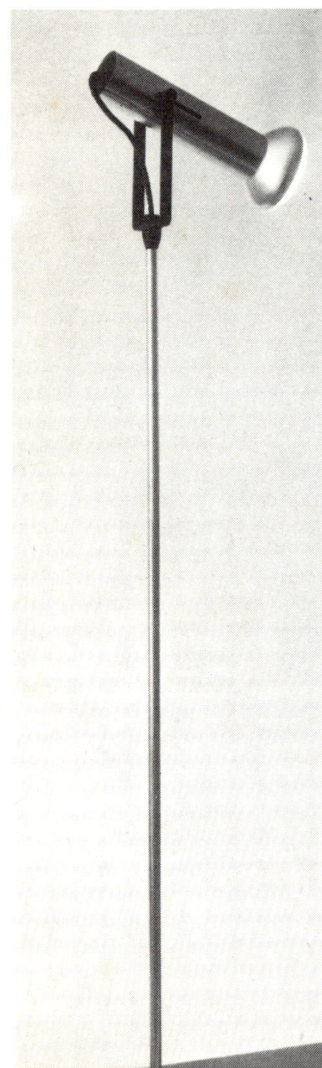

7

8

9

Portable filament

1 Marler Haley
Low Voltage
Spotlight
Compact Spot
12v 50w
incorporating a
transformer

2 Falks
Spotlight Table Lamp
C4469
Holder for 100w
reflector lamp

3 Concord Lighting
International
Spotlight Table Lamp
Quartet Minor
Series 4417/8
100w spotlight in
grey or white (surface
mounted versions
available)

4 Atlas
Retort Stand
Parabolic Spot
Topspot
Parabolic glare-free
spot, adjustable

5 Superswitch Electric
Appliances
Deep Cylindrical
Shade on Table
Standard
LTS 150
Table lamp with
built-in dimmer
control to provide
variable intensity of
light

6 Lumitron
Acrylic Shade Type
3004 or 5
304 mm or 380 mm
diam. acrylic shade

7 Concord Lighting
International
Spotlight on Long
Floor Standard
4231 universal
Floor standard
adjustable spotlight

8 Conelight
Cylinder of Light
KB
Opal plastic shade on
hardwood base

9 Concord Lighting
International
Black Hole Floor
Light
4430
Black hole type of
floor can (upward
light)

Recessed fluorescent

1 Atlas
Recessed Modular Type
1/TMD/2685
Recessed 1500 mm fitting 304 mm wide with plastic diffuser

2 Allom Heffer
Semi-Recessed Modular
707/4/22
Semi-recessed with surface plastic biscuit 608 mm square

Surface mounted fluorescent

3 Concord Lighting International
Opaque Sided Liteframe 7725/7922
1200 mm, 1500 mm, 2400 mm × 304 mm wide, 2 lamps prismatic plate

4 Allom Heffer
Circular Type
707C/2/18
Recessed housing with acrylic diffuser

5 Merchant Adventurers
Timber Sided Opaque
9661/C36
Teak sided with small cell louvre diffuser

6 Atlas
Clip-On Louvre
KU5P
Clip-on louvre/can be attached to any batten fitting to reduce glare

Suspended fluorescent

7 Atlas
Semi-Opaque Sided
A3E/AQ1040
Extruded plastic diffuser 1200 mm

8 Mazda
Kitchen Light Type Netaline
Bare lamp 1200 mm—whilst bare lamps are not recommended they are acceptable in small rooms with limited view. Warm white lamp is supplied —this should be exchanged with supplier for a de-luxe warm white

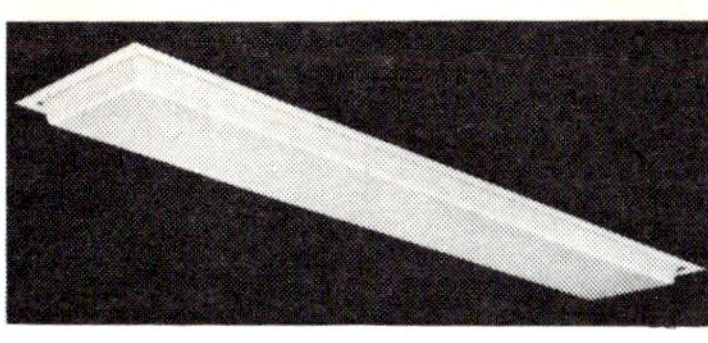

1

2

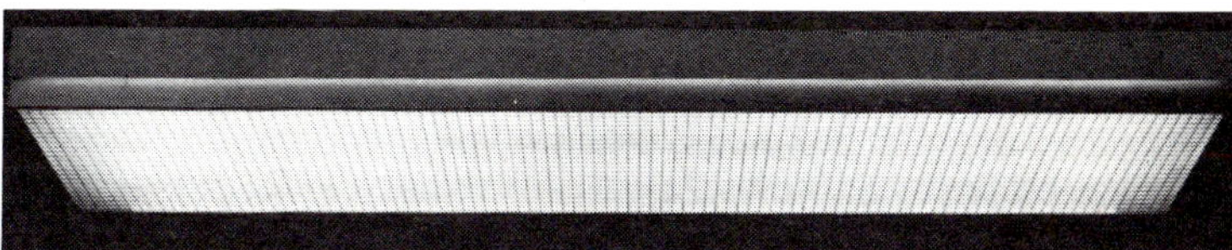

3

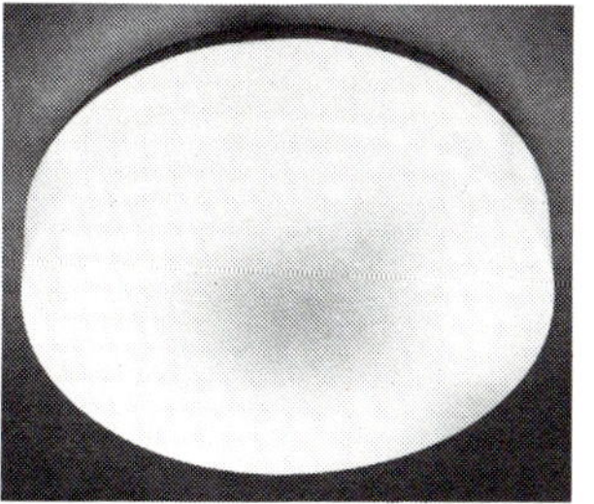

4

5

6

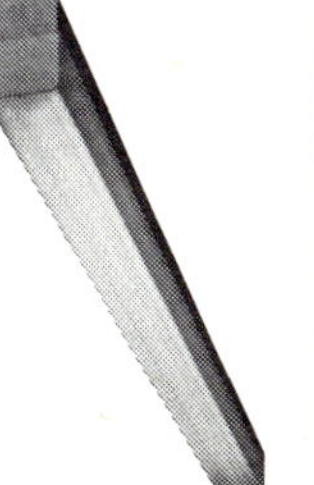

7

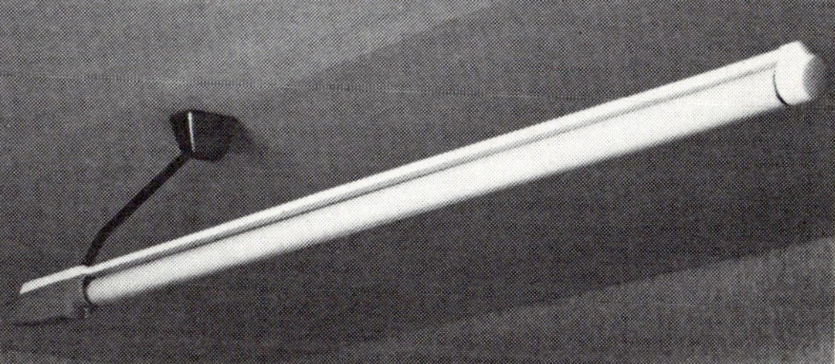

8

Wall-mounted fluorescent

1 Merchant
Adventurers
Picture Light
Light Line 830
Series
Filament version
available for picture
lights 406 mm long.
Aluminium exterior,
up or down light

2 Concord Lighting
International
Bedhead Type
Opaque Front.
Light Up and Down
Silverspan 7980
Series
1200 mm Batten—
opaque front. Insert
in material of one's
own choice to match
decor.

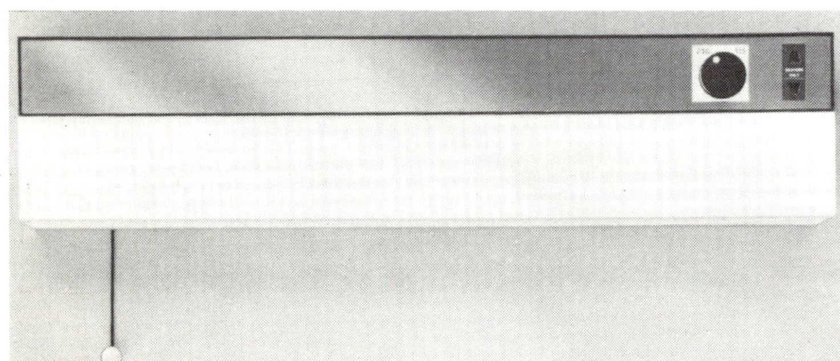

3 Atlas
Mirror Light
View point LST.515
Mirror light with
shaving socket

Portable fluorescent

4 Best & Lloyd
Adjustable Arm
41555
Aluminium tubular
construction 533 mm
13w adjustable

Battens (fluorescent)

Opposite page

1 Atlas Mini batten 456
mm Minipak
Small cross section for
525 mm, 16 mm diam.
lamp

2 Atlas Slim batten and
attachments Arrowslim
Small cross section for
900 mm and 1500 mm
25 mm diam. lamp

3 Osram (GEC) Standard
2 Light Europa range
104 FP S25/28 Twin
tube battens in 1200
mm, 1500 mm, 1800
mm and 2400 mm

4 Atlas Standard Type
  Popular pack 2400 mm
  1800 mm, 1200 mm
  and 600 mm battens

Sections showing typical
uses of batten fittings:

5 Lighting walls or
  curtains

6 Recessed between
  ceiling joists

7 Suspended between
  baffles

8 Behind pelmet

9 Lighting upwards
  behind seat back

10 Lighting ceiling from
  cornice

1

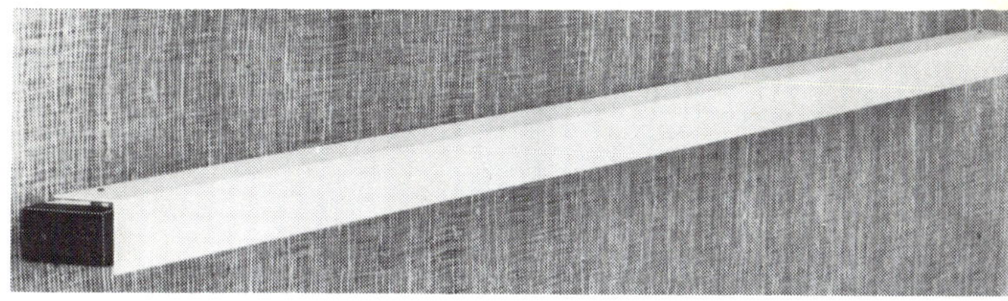

2

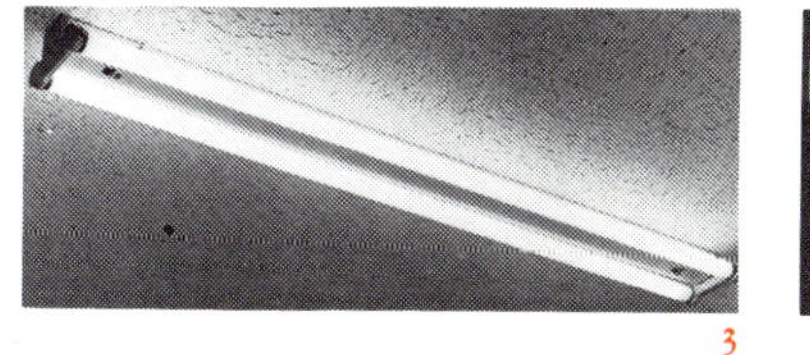

3

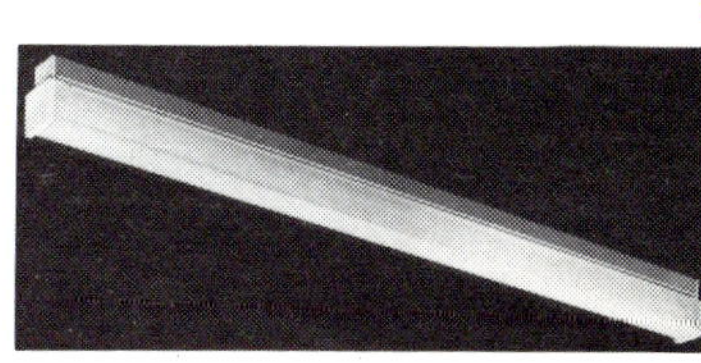

4

Overall lit ceilings (fluorescent)

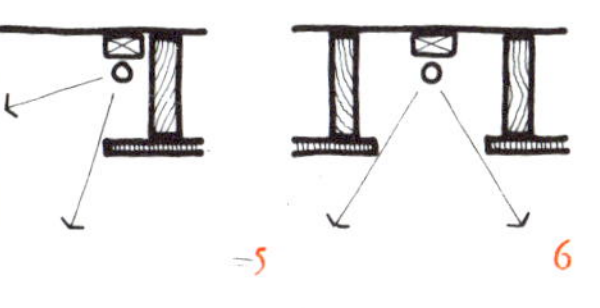
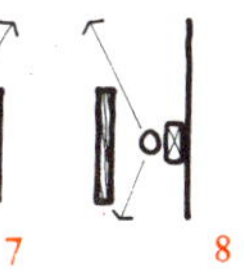
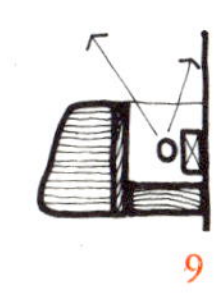

5   6   7   8   9   10

1 Courtney Pope Pantype
  Ceiling Briteglo ceilings
  Vinyl Pan luminous
  ceiling sold as a package
  either 1800 mm ×
  1200 mm or 1800 mm
  × 600 mm

2 Elco Plastics Louvred
  ceiling Elcoplan 9 mm
  cell louvres in 608 mm
  × 608 mm tiles with
  special supporting
  system

3 Isora Metal Louvre
  Leaflite 152 mm deep
  vertical metal fins in
  gold, grey or white

4 Elco Plastics Specular
  Louvre Specular louvres
  Low brightness
  metalised plastic louvres
  directing all light
  downwards

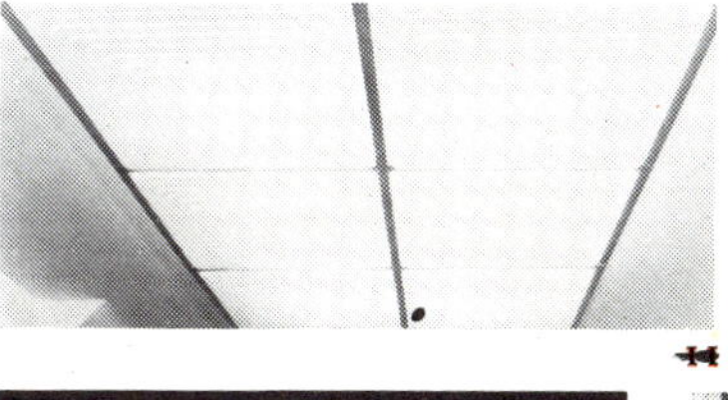

11

12

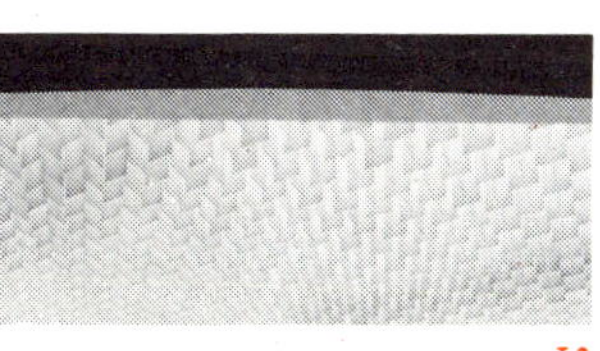

13

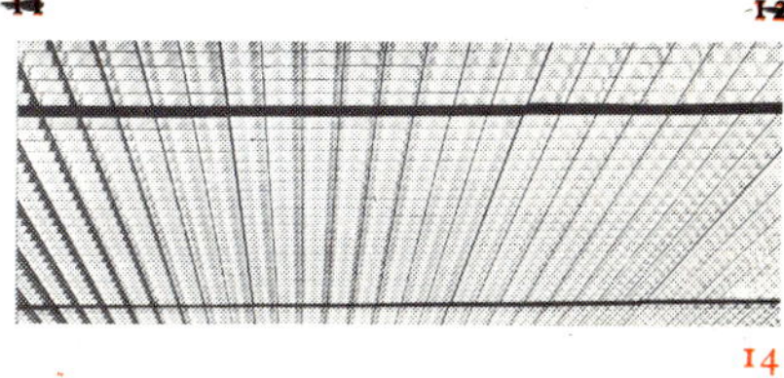

14

Layout of an overall lit
ceiling. The relation-
ship between the spacing
between lamps (S) and
the cavity Depth (D)
is important, for 'even'
appearance over the
luminous ceiling. For
example, with translu-
cent louvres 1·5 to 2D

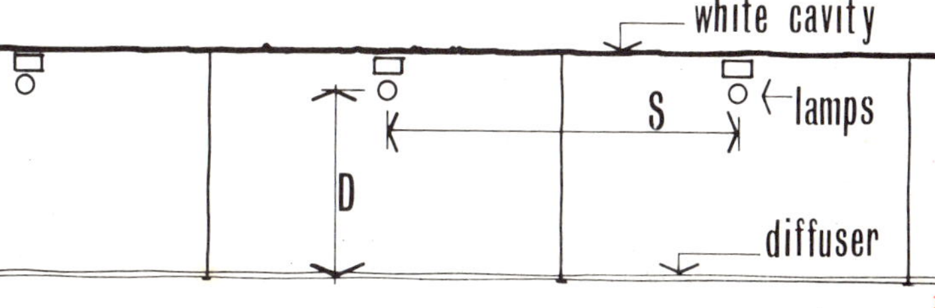

15

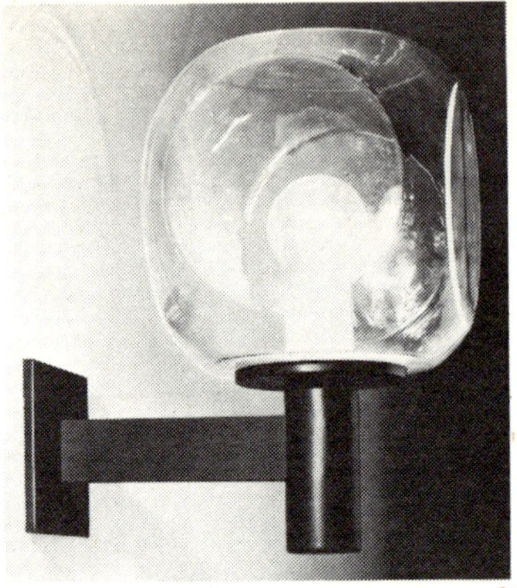

1 **Atlas**
Simple PAR holder
Mini-flood ER1150
Waterproof adjustable
housing for PAR 38
lamp

2 **Merchant
Adventurers**
Wall bracket light
1202 WY7
Smoky or clear glass
on metal wall bracket
for porches

3 **Troughton & Young**
Porch light on soffit
F 916
For use only in
protected situations

4 **Churchouse**
Bulkhead
1128
396 mm bulkhead/
fluorescent 2×300
mm lamp

5 **Concord Lighting
International**
Shield on spike type
Framelite 4230
Louvred shield on a
710 mm rod with
spike; also available
for wall mounting

6 **Churchouse**
Mushroom
9963 Metal
mushroom fitting
Useful for flower bed
lighting

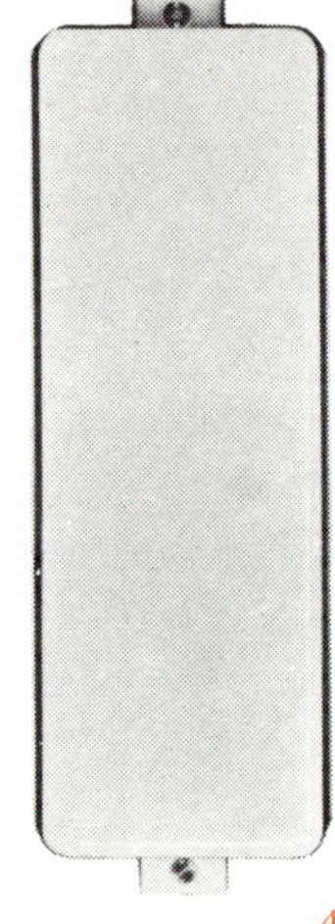

Lamps for domestic use
Type and Outline

	Description and Wattage	Approx. Length	Efficiency lms/w	Special Features	Comments
General Service Filament	Standard 25–200w	105–160 mm	8–14		The most widely used domestic light source. Low lamp cost, no other control equipment required. Warm colour. Simple 'dimming' can be provided. Life 1000 hours. Correct voltage lamp must be used. Up to 1500w available but not generally used in the house.
				Bulb finishes: Clear 40–150w	Has bright filament—recommend only for use with 'sparkle' glass fittings and reflector systems.
				Pearl Pink pearl	Satin finish to bulb obscures direct view of the filament to reduce glare and soften effect.
				Silica coated (60–150w)	Completely diffuses the light from the filament and has uniform brightness.
				Coloured (15–100w)	
				Daylight blue (60–100w)	These are a close match to daylight at a reduction in light of $\frac{1}{2}$ to $\frac{2}{3}$.
					Coiled coil filaments available in 40–150w sizes. Gives 20% higher output than single coil types.
Mushroom	Mushroom 40–150w	89–120 mm	9–13	Silica coated bulb with coiled coil filament. 150w size also available pearl 60–100w pearl	More compact than standard tungsten lamps of similar wattage. Care must be taken to ensure that these compact lamps are not used in fittings designed for lower wattage lamps of standard shape, causing over-heating.
	Three light 40/60/100w 60/100/160w	120–135 mm		Two filaments (40 & 60w) in single bulb enable three light outputs to be achieved by switching. Screw cap for special holder (alternatively 60/100)	Recommended for use where it is desirable to change the lighting level at certain times.

Lamps for domestic use
Type and Outline

	Description and Wattage	Approx. Length	Efficiency lms/w	Special Features	Comments
Special Filament Bulb Shapes 1. Fantasie	1 Fantasie 40w	133 mm		One shape at present available. Silica coated	The increased size and completely diffusing bulb is designed as a complete fitting, but care must be taken to prevent glare.
	2,3,4 Candle 25–60w	92–128 mm		Clear, pearl, coloured or silica coated	Designed to allow a number of smaller light sources in a fitting. Not to be confused with the mobility and sparkle of a real candle.
	5 Pigmy 15–25w	56 mm		Clear, pearl or coloured	Can be used in very confined spaces.
	6 45 mm round 25–40w	65 mm		Silica coated 15–25w coloured	
	Pilot 6–10w	48 mm		Clear	Similar to pigmy lamps.
Filament Reflector	1 Internally silvered 100–150w	135–178 mm	10–11	Spotlamp Beam 40 deg.–45 deg. Floodlamp Beam 90 deg.–100 deg.	Integral reflector provides a controlled beam of light without additional reflector, useful for projecting light onto walls and spotlighting objects. Life 1000 hours.
	2 PAR 38 Pressed Glass 100–150w	135 mm	12	Spotlamp Beam 25 deg.–30 deg. Floodlamp Beam 60 deg.–65 deg.	More expensive than standard internally silvered lamps but with a life of 1500 hours. The beam is more controlled than with internally silvered types. Red, blue, green, yellow 100w also.
	3 Coated Mushroom 60w	93 mm			A mushroom lamp in which the top has been internally silvered to project light, gives a direct illumination of 30% more than standard mushroom lamp.
	4 Crown silvered 100w	120 mm	10		Used in association with a reflector—the silvered bowl cutting out the direct view of the filament from below.

Lamps for domestic use
Type and Outline

		Description and Wattage	Approx. Length	Efficiency lms/w	Special Features	Comments
Tubular Filament	1	Striplite 30–60w	221–284 mm		Clear or opal	Tubular filament lamps are still widely used either concealed or in association with metal reflectors, e.g. picture lights. The complete fitting will have a lower initial cost than fluorescent lighting but with its life of 1000 hours its lamp replacement cost will be higher.
	2	Single end tubular 25–60w	304 mm		Opal	
					Clear	
	3	Architectural straight 35–150w	304–1220 mm		Opal	All tubular lamps are less efficient than general service types. They produce more heat for a given amount of light but useful where fluorescent would give too much light.
	4	Curved $\frac{1}{8},\frac{1}{4},\frac{1}{2}$ circles 60w	Curved length 500 mm			
		Maxstrip 40–60w	252 mm		Clear and opal	
Low Voltage Filament	1	Reflector 24 & 50w 12v	70–82 mm	12	Beam 24w = 30 deg. Beam 50w = 20 deg. Beam 20w = 30 deg.	Require transformer at extra cost for operation from normal mains supply.
		20w 24/28v	70 mm			Very compact source, suitable for small reading lamps and spotlighting small objects.
	2	Crown silvered 50w 12v 100w 24v 150w 24v	70 mm 120 mm 82 mm	12		Produces very closely controlled sharp edge beams when associated with a parabolic reflector. Beam angles as small as 8 deg. obtainable.
	3	Display Tungsten Halogen		20–22	12v 50w	High efficiency. 2000 hr. life.
Neon Types	1	Neon 0·5w	56 mm			Used as an indicator in switches and equipment.
	2	Nightlight 8w	105 mm			Ideal for providing a maintained low level of lighting at night.

Lamps for domestic use
Type and Outline

Description and Wattage	Approx. Length	Efficiency lms/w	Special Features	Comments
Tubular Fluorescent				
1 Straight 15w to 125w	450–2400 mm 38 mm diam. 450–900 mm, 2400 mm in 25 mm diam.	20–60 depending on length & colour	Wide range of colours available. High efficiency types give a colour of light generally considered unsuitable for domestic use, and the lower efficiency de luxe colours are recommended for the home. (5 ft & 4 ft available in red, green, yellow, blue, pink & gold for decorative use.)	Control equipment must always be used with fluorescent and is normally housed within the lighting fitting. In certain cases however, where space is restricted, the control gear may be placed some distance from the lamp. All control gear consumes a small amount of electrical power and this has been taken into account in the efficiencies stated. Normal life 5000 hours. Dimming fluorescent lamps requires relatively expensive equipment. Some fluorescent lamps may be operated without control gear by using a 'ballast circuit'. In this circuit a special filament or 'ballast' lamp is used.
2 Miniature 4–13w	150–525 mm 15 mm diam.	23–38	Very limited, colour range	Small diameter permits concealment in very confined spaces.
3 Circular 22–40w	210–406 mm diam. of circle 38 mm diam. of tube	23–44		These lamps overcome the linear characteristic of normal fluorescent lamps and make them more acceptable for fittings at the domestic scale.
Tubular Fluorescent Reflector				
4 Reflector 20–125w	600–2400 mm	35–48	Not widely produced in the de luxe colours	A built-in reflector is used to increase the directional light output.
5 Straight up to 91w	up to 2900 mm lighting length	20–36	Any shorter length can be made. Tubes can be preformed into any special shape	Slightly lower efficiency than normal fluorescent but the life is 15000 hours.
Cold Cathode Fluorescent				
Hairpin 67–91w	1440 mm	20–36	High efficiency and some de luxe colours available	Require transformers for operation. Useful for following curved architectural forms. By bending the electrodes back continuous lines of light can be produced. Simple dimming methods are available. Lamps operate at high voltage.

Appendix (a) Costs

For home lighting the costs can be analysed as follows:

Capital Costs   1. Electrical Installation
                     2. Lighting Fittings

Running Costs 3. Power
                     4. Lamp Replacement.

For built-in lighting there may be an additional cost in alteration or modification to structure for the incorporation of lamps and lighting equipment.

Capital Costs

The installation costs for lighting are difficult to assess first because it is usual for the lighting and power points to be priced together and power points are, of course, used for other electrical needs besides lighting. Second, the price of light fittings varies greatly according to the quality of equipment used.

As a guide, however, the wiring for the average house costs £60* providing for ten lighting points and seven power points, whilst the wiring to the improved standards would cost more like £115 (providing 20 lighting points and 20 power points). No realistic comparison can be made between the costs of equipment when it is realised that it is possible to obtain adequate fittings for as little as £1 each or as much as £10 or £15, but a good average might appear to be £4 per unit. At this price the total fittings expenditure would be £80. Total capital expenditure is, therefore, £115 + £80 = £195.
Total Capital Cost amortised over a ten year period (say 10% p.a.) = 7s 6d a week.

Running Costs

The average number of lighting points per dwelling in 1965 was quoted as just under ten. **Taking this number, the average cost of

---

*Prices based on new housing schemes where a number of similar dwellings are being built. For individual sites the price would be higher. Prices for additions to wiring will, of course, be higher and it is advisable to provide more than may appear necessary at the start, to allow for future needs.

**Electricity Council.

lighting for the year at a unit cost of electricity of 2*d* would be at the most £4 15*s* 0*d*. The cost of lamp replacement would be 15*s* approximately. Total running costs of £5 10*s* 0*d* per year of approximately 2*s* 1*d* a week average over the year. The costs of running a lighting installation to the improved standards outlined for a similar sized house might total £21 0*s*. 0*d* or 8*s* per week.

Total Costs	Capital Costs	7*s* 6*d*
	Running Costs	8*s* 0*d*
		———
		15*s* 6*d* per week
		———

Appendix (b) Wiring and Switching

Wiring

The provision of an adequate wiring system is the basis for effective lighting; it contributes to safety in the home not least by eliminating the need to run a number of electrical appliances or portable lighting fittings from one socket outlet. When cuts in building costs have to be made it is easy to reduce the electrical specification for a house, but this should be resisted in the interests of functional living.

It would not be advisable to specify here how many lighting points should be provided in each room, since the requirements of different house plans are unique. If the general lighting principle of 'full information' is adopted it should be possible to plan for any optimum number of lighting points, either for permanently fixed units, or for the addition of portable equipment.

Ring Circuit

This method is now becoming an accepted standard for all new houses, and may be applied to old properties when the wiring is dangerous or inadequate. The system permits any number of 13 amp fused socket outlets to be placed in an area of 93 sq. m. For areas larger than this, two ring mains will generally be installed. This circuit is not normally used for fixed lighting points but for socket outlets from which portable lighting fittings may be run.

The principle by which this operates is illustrated here, the required number of socket outlets being placed on the ring at strategic positions in rooms or circulation areas. When socket outlets are required remote from the most economic path of the ring of cable, spurs may be taken off, but these must be limited to two sockets.

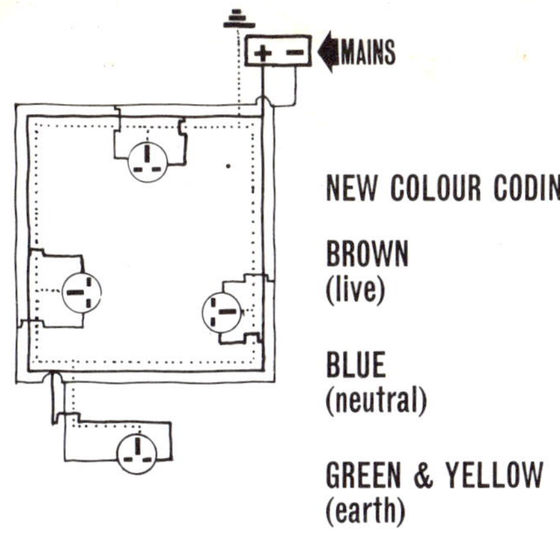

Ring circuit
Note: one ring circuit
must be installed
for every 93 sq. m.
of floor space

Special Lighting and Power Distributions

(a) Lytespan track available from Concord Lighting International Ltd consists of a special extruded aluminium section mounted either in or on the ceiling or wall, which contains a small section electrical 'bus-bar' to which fittings with a special adapter can be plugged in anywhere along its length. The standard length is 2400 mm or 1200 mm but lengths can be joined to form any desired pattern.

(b) In America a 110 volt strip-plug is available which can be placed at skirting board, or window cill level, around a room enabling plugs to be placed anywhere along its length, shortening the distance between socket and lighting fitting, thus avoiding the danger of trailing flexes and overloaded sockets. Strip-plug is not yet available in Great Britain for 220-240 volts.

Concord Lytespan track
The use of Lytespan track to enable light fittings to be placed at any position along its length. (Recessed version)

(c) A simple method of providing adjustibility of position of pendant light fittings is by providing 'Silent Gliss' curtain track and attaching the cable to it in loops so that a light fitting can be pulled across a room.

(d) A further simple device is available from Troughton and Young which consists of a spreader attachment for the normal ceiling rose enabling two or more fittings to be taken from a single point. Each fitting is then supported by a small ceiling pin. Care must be taken to ensure that the combined wattage is not too great for the single point.

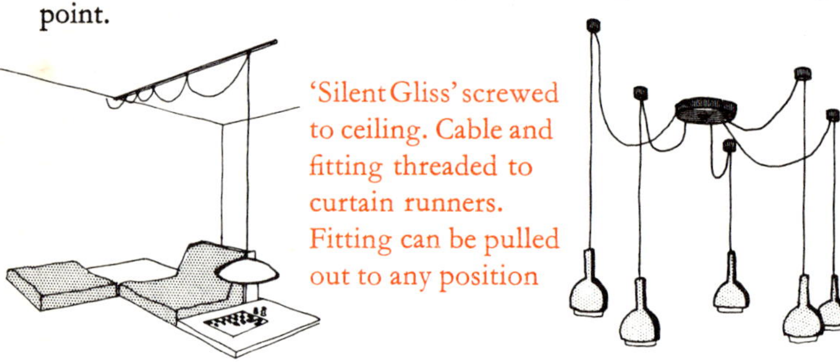

'Silent Gliss' screwed to ceiling. Cable and fitting threaded to curtain runners. Fitting can be pulled out to any position

Distribution from a single point. Troughton & Young.
Also available: Concord, Atlas, Merchant Adventurers

Switching

Switching can be either 'single way' or 'two way'. Single way switching is used when a single switch position is adequate. It can control more than one light.

Two way switching should be planned for circulation areas, corridors and stairs, garages with internal access to houses, or wherever it is advantageous for the light to be switched on and off from more than one place. Two way switches can operate one or more points.

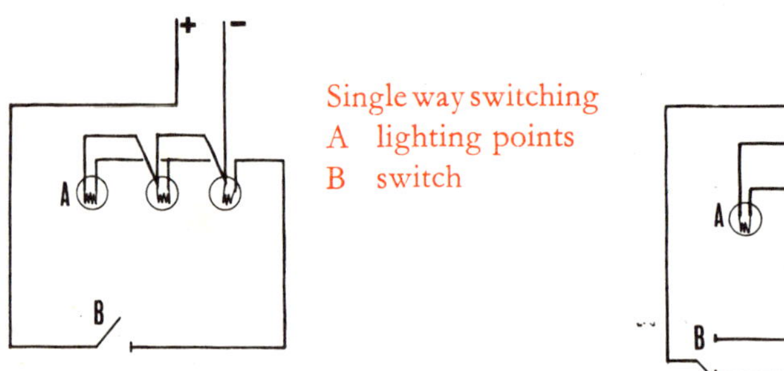

Single way switching
A   lighting points
B   switch

Two way switching
A   Lighting points
B   2 way switch
Shown off. Moving either switch will complete circuit

Switches

There are two main types   1. Toggle
                           2. Rocker type

The Rocker type tend to look better and have the advantage of being switched easily with arm or elbow.